Constructive Aspects
of the
Fundamental Theorem
of
Algebra

Constructive Aspects
of the
Fundamental Theorem
of
Algebra

Proceedings of a Symposium
Conducted at the
IBM Research Laboratory, Zürich-
Rüschlikon, Switzerland, June 5–7, 1967

edited by

Bruno Dejon
IBM Research Laboratory, Zürich

Peter Henrici
Professor of Mathematics, Eidgenössische Technische Hochschule Zürich

WILEY-INTERSCIENCE
A division of John Wiley & Sons Ltd.
London · New York · Sydney · Toronto

Copyright © 1969 John Wiley & Sons Ltd. All rights reserved. No part of this book may be reproduced by any means, nor transmitted, nor translated into a machine language without the written permission of the publisher.

Library of Congress catalog card number 69–19830

SBN 471 20300 9

Printed in Great Britain by J. W. Arrowsmith Ltd., Bristol

Preface

A symposium entitled "Constructive Aspects of the Fundamental Theorem of Algebra" was held on June 5 - 7, 1967, at the IBM Research Laboratory in Rüschlikon near Zürich (Switzerland). The present volume contains all invited papers and most of the short communications presented at that symposium.

The purpose of the symposium was to gather some experts from the fields of constructive and of numerical analysis, and to let them discuss the significance of the concept of constructivity in the context of a classical and well-defined problem of numerical computation, namely the problem of determining the zeros of a polynomial. It is clear that due to obvious limitations only a small selection of all possible points of view could be represented at the symposium. Those who were present will nevertheless agree that the discussions were informative and stimulating. It is hoped that the publication of these proceedings will contribute further to the understanding between the theoreticians and the practicioners of computation.

We are indebted to the IBM Corporation for their support of the symposium, to the speakers for contributing their manuscripts, and to Miss E. Enzler for preparing the manuscripts for publication.

Rüschlikon, September 1968   B. Dejon   P. Henrici

Contents

Preface	v
DEJON B., NICKEL K.: A Never Failing, Fast Convergent Root-Finding Algorithm*	1
DEKKER T.J.: Finding a Zero by Means of Successive Linear Interpolation	37
FORSYTHE G.E.: Remarks on the Paper by Dekker	49
FORSYTHE G.E.: What is a Satisfactory Quadratic Equation Solver?	53
FOX L.: Mathematical and Physical Polynomials	63
GOODSTEIN R.L.: A Constructive Form of the Second Gauss Proof of the Fundamental Theorem of Algebra*	69
HENRICI P., GARGANTINI I.: Uniformly Convergent Algorithms for the Simultaneous Approximation of all Zeros of a Polynomial*	77
HERMES H.: On the Notion of Constructivity*	115
HOUSEHOLDER A.S., STEWART G.W., III: Bigradients, Hankel Determinants, and the Padé Table*	131
JENKINS M.A., TRAUB J.F.: An Algorithm for an Automatic General Polynomial Solver*	151
KUPKA I.: Die numerische Bestimmung mehrfacher und nahe benachbarter Polynomnullstellen nach einem verbesserten Bernoulli-Verfahren	181
LEHMER D.H.: Search Procedures for Polynomial Equation Solving*	193

OSTROWSKI A.M.: A Method for Automatic Solution of Algebraic
    Equations* 209
PAVEL-PARVU M., KORGANOFF A.: Iteration Functions for Solv-
    ing Polynomial Matrix Equations* 225
RUTISHAUSER H.: Zur Problematik der Nullstellenbestimmung
    bei Polynomen* 281
SCHRÖDER J.: Factorization of Polynomials by Generalized
    Newton Procedures 295
SPECKER E.: The Fundamental Theorem of Algebra in Recursive
    Analysis* 321
Index 331

---

* Invited paper

Bruno Dejon and Karl Nickel*

A Never Failing, Fast Convergent Root-Finding Algorithm

Summary

      A new method for the computation of all roots of an arbitrary polynomial is presented. A complete PL/I procedure is given and the results of some thousand examples are discussed. The convergence of the method is proven with the aid of a widely applicable abstract convergence theorem.

1. Introduction

      In a previous paper [3] a method has been presented for the computation of all roots of an arbitrary given polynomial $P(z)$. This method included the evaluation of error bounds for the computed root approximations. For this reason a certain kind of round-off error estimation had to be used during the arithmetic operations ("Fehlerschranken-Arithmetik", see [2], Moore's "interval-arithmetic", see [1] or Triplex ALGOL 60, see [6]). This new arithmetic was also used in decisions for stopping a cycle during the computation.

      There is, however, a general lack of such an arithmetic in common computer languages such as FORTRAN, ALGOL or PL/I. Therefore, in what follows our method is changed in such a way that no

---

*)   This is a slightly altered version of the lecture presented by K. Nickel at the Symposium.

"Fehlerschranken-Arithmetik" has to be used. The iteration is stopped if either $P(z^*) = 0$ for some complex number z^*, or if the z-increment is too small. As round-off errors cannot be taken under control, no error bound for the computed approximate root z^* can be given.

2. Description of the method

Let $n > 0$ be an integer and

(1) $$P(z) := \sum_{j=0}^{n} a_j z^j, \quad z := x+iy .$$

Our algorithm is an iterative one.

We choose as an initial approximation

(2) $$z^* := -a_{n-1}/na_n .$$

This is the "middle point" of the roots w_p (see [3]) and the exact solution for $n = 1$. Let us state, however, that the convergence of the method will not depend on the choice of the initial approximation.

We then improve the approximation z^* by applying the transformation T defined hereafter. It consists in trying first a so-called *normal step* A (providing for an improvement of the given approximation like with a quadratically convergent method). If step A fails (in a sense to be defined later) we proceed to a *convergence enforcing emergency step* B which ensures linear convergence.

A d s t e p A :
For a given z^* let

(3) $$c_j = c_j(z^*) := P^{(j)}(z^*)/j! \quad \text{for } j = 0(1)n.$$

Then the identity

$$P(z) = \sum_{j=0}^{n} c_j (z-z^*)^j$$

holds.

Now set

(4) $$\hat{z} := z^* + (-c_0/c_k)^{1/k}$$

where $k \in \{1(1)n\}$ is such that $c_k \neq 0$ and

(5) $$r := |c_0/c_k|^{1/k} = \min_{j=1(1)n} |c_0/c_j|^{1/j}.$$

If several such k exist, take the smallest one. The branches of the multivalued analytic functions $(...)^{1/j}$, $j = 2(1)n$, occurring in (4) and (5) are chosen such that $z^{1/j}$ is positive for z positive.

Let ε be a fixed real number with $0 < \varepsilon < 1$.

If

(6) $$|P(\hat{z})| \leq (1-\varepsilon)|P(z^*)|$$

then set

(7) $$Tz^* := \hat{z}.$$

If (6) does not hold, proceed to step B.

Ad step B.

We first introduce transformations T_m, $m = 1, 2, \ldots$, by defining

(8) $$T_m z^* := z^* + 2^{-m} r \, (-c_0/c_\ell \cdot |c_\ell/c_0|)^{1/\ell},$$

where the index $\ell = \ell(m) \varepsilon \{1(1)n\}$ is chosen such that $c_\ell \neq 0$ and

(9) $$|c_\ell| 2^{-m\ell} r^\ell = \max_{j=1(1)n} |c_j| 2^{-mj} r^j .$$

If several such ℓ exist, take the smallest one.

Let $\bar{m}$ be the smallest index m such that

(10) $$|P(T_m z^*)| \leq (1 - 2^{-m\ell - 1} r^\ell |c_\ell/c_0|) \, |P(z^*)|.$$

We then set

(11) $$T z^* := T_{\bar{m}} z^*.$$

To verify that such a natural number $\bar{m}$ exists, we first observe that the inequalities

(12) $$|c_\kappa| 2^{-m\kappa} r^\kappa > 2 \sum_{\substack{j=1 \\ j \neq \kappa}}^{n} |c_j| 2^{-mj} r^j, \quad \kappa \varepsilon \{1(1)n\},$$

valid for some positive integer m, imply (by the defining relation (9)) that

$$\ell(m) = \kappa.$$

Furthermore

$$|P(T_m z^*)| \leq |c_0 + c_\kappa (T_m z^* - z^*)^\kappa|$$

$$+ \Big| \sum_{\substack{j=1 \\ j \neq \kappa}}^{n} c_j (T_m z^* - z^*)^j \Big|$$

$$\leq |c_0|(1 - 2^{-m\kappa} r^\kappa |c_\kappa/c_0|)$$

$$+ \sum_{\substack{j=1 \\ j \neq \kappa}}^{n} |c_j| 2^{-mj} r^j$$

$$\leq |c_0|(1 - 2^{-m\kappa - 1} r^\kappa |c_\kappa/c_0|),$$

i.e. inequalities (12) imply inequality (10).

Let us assume κ is such that

$$P'(z^*) = P''(z^*) = \ldots = P^{(\kappa-1)}(z^*) = 0,$$

$$P^{(\kappa)}(z^*) \neq 0.$$

Then one easily sees that there exists a natural number M such that (12) and, hence (10) hold with $m = M$.

Therefore the set of natural numbers m fulfilling inequality (10) is not empty, whence, a minimal index $\bar{m}$ exists.

Observation: The convergence proof of Section 6 will be independent of the special form chosen for the normal step. All we shall use will be the fact that there exists a constant $\varepsilon > 0$ such that

$$|P(Tz^*)| \leq (1-\varepsilon)|P(z^*)|,$$

whenever T is achieved by the normal step.

3. Computer implementation of the algorithm

The computer program of our algorithm as presented below contains a normal step and an emergency step.

In the normal step formulas (3) to (7) are programmed, with the difference, however, that inequality (6) is programmed as a test simply for

(13) $$|P(\hat{z})| < |P(z^*)|.$$

As there is a smallest positive machine number, say N_1, and a largest one, say N_2, we are still sure that there exists an $\varepsilon > 0$,

$$\varepsilon = N_1/N_2$$

for instance, such that

$$|P(\hat{z})| \leq (1-\varepsilon)|P(z^*)|$$

whenever test (13) is passed successfully.

The emergency step of the program reflects formulas (8) and (9), while inequality (10), similarly to what was done in the normal step, is replaced by a test for

(14) $$|P(T_m z^*)| < |P(z^*)|.$$

Here again, however, we are sure that there exists an $\varepsilon > 0$, $\varepsilon = N_1/N_2$ for instance, such that

$$|P(T_m z^*)| \le (1-\varepsilon)|P(z^*)|,$$

whenever test (14) is passed successfully.

The probably most conspicuous feature that distinguishes a computer algorithm from a theoretical one is the need for stopping conditions.

In the program below there are two states of the computer that lead to a stop. The first one is reached when z^* is a "numerical root", i.e. when $P(z^*) = 0$ within floating point numbers accuracy. The second terminating state occurs when $Tz^* - z^*$ is numerically zero by underflow, i.e. if a further iteration is useless.

The following program is a PL/I-version of our algorithm.

Title   : Subroutine POLROOT (N,A,W) for the root-finding for polynomials

Type   : PL/I procedure

Purpose: Given the integer degree N and the complex coefficients A(I) with A(N) ≠ 0 of the polynomial

$$P(Z) := \sum_{I=0}^{N} A(I)\ Z^I.$$

The subroutine POLROOT computes the N roots W(I) such that

$$P(Z) = A(N) \prod_{I=1}^{N} (Z-W(I)) \ .$$

Procedure: The algorithm given in Numerische Mathematik *9*, 80-98 (1966) is changed in such a way that no "Fehlerschranken-Arithmetik" has to be used. The iteration is now stopped if either P(W(I))=0 or if the Z-increment is too small.

Notes :
1. The degree N has the characteristics: binary, fixed-point, real with the precision 15.
2. The complex coefficients A(I) for I=0(1)N and the computed roots W(I) for I=1(1)N have the characteristics: decimal, floating-point, complex with the precision 16.
3. If, during the computation, too large numbers occur (e.g. A(N) too small compared with the other coefficients A(I)), the following message will be printed: "POLROOT-ERROR, COEFFICIENTS TOO LARGE" and the computer returns to the main program.
4. The subroutine is not affected by multiple roots or clusters of roots, but normally in these cases the errors of the computed roots are higher than in the case of simple roots, and more time is required for the computation.

```
   (NOUNDERFLOW) :
POLROOT : PROCEDURE (N,A,W) ;
   DECLARE (I,J,K,L,M,N) BINARY FIXED REAL (15) ;
   DECLARE (ADO,ADK,ADI,EINS,MA,MAX,R) DECIMAL FLOAT REAL (16) ;
   DECLARE (Q,Y,Z) DECIMAL FLOAT COMPLEX (16) ;
   DECLARE (A,B,C,D,W) (0:N) DECIMAL FLOAT COMPLEX (16) ;
   ON OVERFLOW BEGIN ;
               PUT SKIP ;
               PUT LIST ('POLROOT-ERROR,COEFFICIENTS TOO LARGE') ;
               PUT SKIP ;
               GOTO OMEGA ;
               END ;
   EINS = 1 ;
   M = N ;
   B = A ;
   E1 : IF M = 0 THEN GOTO OMEGA ;
        Z = - B(M-1)/(M*B(M)) ;
        IF M = 1 THEN   G1 : BEGIN ;
                             Y = Z ;
                             GOTO E6 ;
                             END G1 ;
   E2 : D = B ;
        F1 : DO J = 0 TO M-1 ;
             F2 : DO I = M-1 TO J BY -1 ;
                  D(I) = D(I) + Z*D(I+1) ;
                  END F2 ;
        END F1 ;
        IF D(0) = 0 THEN GOTO G1 ;
        K = M ;
        ADO = ABS (D(0) ) ;
        ADK = ABS (D(M) ) ;
        R = ADO**(EINS/M) / (ADK**(EINS/M) ) ;
        F3 : DO I = M-1 TO 1 BY -1 ;
             ADI = ABS ( D(I) ) ;
             IF ADI <= ADO**(FLOAT(K-I,16)/K) * ADK**(FLOAT(I,16)/K)
                     THEN GOTO E3 ;
             R = ADO**(EINS/I) / (ADI**(EINS/I) ) ;
             K = I ;
             ADK = ADI ;
   E3 : END F3 ;
        Y = Z + (-D(0))**(EINS/K)/((D(K))**(EINS/K)) ;
        IF Z = Y THEN GOTO E6 ;
        C(M) = B(M) ;
        F4 : DO I = M-1 TO 0 BY -1 ;
             C(I) = B(I) + Y*C(I+1) ;
             END F4 ;
```

```
              IF ABS(C(0)) < ADO THEN G2 : BEGIN ;
                                           Z = Y ;
                                           GOTO E2 ;
                                     END G2 ;
   E4 : MAX = 0 ;
        R = R/2 ;
        F5 : DO I = 1 TO K ;
             MA = ABS(D(I))*(R**I) ;
             IF MA < MAX THEN GOTO E5 ;
             MAX = MA ;
             L = I ;
   E5 : END F5 ;
        Q = -D(0)/D(L) ;
        IF Q = 0 THEN GOTO E6 ;
        Y = Z + R * (Q/ABS(Q))**(EINS/L) ;
        IF Y = Z THEN GOTO E6 ;
        C(M) = B(M) ;
        F6 : DO I = M-1 TO 0 BY -1 ;
             C(I) = B(I) + Y*C(I+1) ;
        END F6 ;
        IF ABS(C(0)) < ADO THEN GOTO G2 ;
        K = L ;
        GOTO E4 ;
   E6 : W(M) = Y ;
        F7 : DO I = M-1 TO 0 BY -1 ;
             B(I) = B(I) + Y*B(I+1) ;
        END F7 ;
        M = M-1 ;
        F8 : DO I = 0 TO M ;
             B(I) = B(I+1) ;
        END F8 ;
        GOTO E1 ;
OMEGA : ;
END POLROOT ;


Test-program:
N=5;A(0)=1;A(1)=A(2)=A(3)=A(4)=0;A(5)=1
and
N=5;A(0)=10^{60};A(1)=A(2)=A(3)=A(4)=0;A(5)=10^{-60}
```

```
TEST : PROCEDURE OPTIONS (MAIN) ;
   DECLARE (I,N) BINARY FIXED REAL (15) ;
   DECLARE (A,W) (0:5) DECIMAL FLOAT COMPLEX (16) ;
   N = 5 ;
   A = 0 ;
   A(N) = 1 ;
   A(0) = 1 ;
   PUT PAGE ;
   CALL POLROOT (N,A,W) ;
   PUT LIST (N) SKIP ;
   PUT SKIP ;
 R: DO I = 0 TO N ;
   PUT LIST (A(I)) SKIP ;
   END R ;
   PUT SKIP ;
 S: DO I = 1 TO N ;
   PUT LIST (W(I)) SKIP ;
   END S ;
   A(N) = 1.0E-60 ;
   A(0) = 1.0E+60 ;
   PUT PAGE ;
   CALL POLROOT (N,A,W) ;
   PUT LIST (N) SKIP ;
   PUT SKIP ;
 T: DO I = 0 TO N ;
   PUT LIST (A(I)) SKIP ;
   END T ;

      Procedure POLROOT

END TEST ;
```

4. Examples

The subroutine POLROOT was tested on the IBM System/ 360 model 40 computer at the IBM Zurich Research Laboratory. During these tests more than 10,000 polynomial roots were calculated in the following way (see [3]): after choosing the degree n, n complex values w_p were calculated as random numbers. Then the coefficients a_j of the polynomial

$$P(z) := \sum_{j=0}^{n} a_j z^j$$

were computed according to the indentity

$$P(z) = a_n \prod_{p=1}^{n} (z - w_p) .$$

Finally, approximate values $\hat{z}$ for the exact roots w_p were evaluated with the aid of subroutine POLROOT.

Thus, both values $\hat{z}$ and w_p were available, the absolute and the relative errors $|\hat{z} - w_p|$ and $|\hat{z}-w_p|/|w_p|$ could be computed and a statistical analysis of the errors made.

Some typical results for simple roots are given in Table I. The roots w_p were generated in the square $|Re\ z| \leq 2$, $|Im\ z| \leq 2$, using a subroutine "RANDOM" $r[\ ]$ of Professor Rutishauser ([4], with initial value $z=1$ and parameters $p=1$, $q=2$). The formula used was

$$Re\ w_p := 2\ r[1] - 4, \quad Im\ w_p := 2\ r[2] - 4 \qquad p = 1(1)n .$$

In Table II, for a polynomial of degree $n = 100$ (!), the values of w_p for $p = 1(1)100$, of a_j for $j = 0(1)100$ and of some interesting values are reproduced. The geometric distribution of the random numbers w_p in the square $|Re\ z| \leq 2$, $|Im\ z| \leq 2$ is shown in Fig. 1.

As may be seen from Table II the first roots evaluated by POLROOT are the ones with smallest modulus. The sequence of the $\hat{z}$ has approximately monotonic-increasing absolute value. The worst relative error occurred in the 92nd evaluation with a value of $\approx 1_{10}-6$. The corresponding root w_{21} is very close to w_{42} which exhibits the second worst value, $\approx 0.97_{10}-7$, of the relative error.

The next three roots for which the approximation is bad are w_3, w_{71} and w_{85} which form (see Fig. 1) a cluster of roots. On the other hand, a great number of roots is evaluated with very high accuracy, e.g. w_{52}, w_{38}, w_6, w_{95} and w_{98}, all with a relative error $< 1_{10}{-14}$.

Also the case of multiple roots has been treated. A typical example is given in Table III. Here the degree is non-random, $n = 10$ in all examples. The roots w_p are chosen as random numbers such that q roots lie inside the small square $|Re\ z - 1| \leq 0.01$, $|Im\ z| \leq 0.01$; the remaining $n - q$ roots are distributed in the larger square $|Re\ z| \leq 2$, $|Im\ z| \leq 2$. The formula used was

$$\left. \begin{array}{l} Re\ w_p := 0.995 + 0.01\ r[1], \\ \\ Im\ w_p := 0.01\ r[2] - 0.005 \end{array} \right\} \text{ for } p = 1(1)q\ ;$$

$$\left. \begin{array}{l} Re\ w_p := 2\ r[1] - 4\ , \\ \\ Im\ w_p := 2\ r[2] - 4 \end{array} \right\} \text{ for } p = q+1(1)n\ .$$

The accuracy of the approximations $\hat{z}$ decreases sharply with increasing "multiplicty" q as can be foreseen by an error analysis (see Wilkinson[5]).

Finally, the coefficients c_j were generated as (real or complex) random numbers. In this case the exact roots are unknown and an error statistic is therefore impossible. Because of the similarity of the results to those given in [3] a reproduction is omitted.

Figure 2 represents the results of another problem: While the function $\exp z := \sum_{j=0}^{\infty} z^j/j!$ has no zeros in the whole complex plane, the partial sums $\sum_{j=0}^{n} z^j/j!$ are polynomials of degree n and, therefore have exactly n roots w_p, $p = 1(1)n$. How do these roots w_p behave for $n \to \infty$? The results for $n = 1(1)47$ were computed with the subroutine POLROOT and are shown in Fig. 2. All roots with the same number n are connected by a line.

5. An abstract convergence theorem for monotonic algorithms

<u>Theorem</u>: Let $B := \{u, v, \ldots\}$ be a compact metric space. Let the real-valued function $a(u) \geq 0$ be defined and continuous in B with exactly $\bar{n}$ different zeros $u_p \epsilon B$, $p = 1(1)\bar{n}$. Let T be a transformation which maps B into B and has the following properties:

    a) T is continuous at the $\bar{n}$ zeros u_p of $a(u)$.

    b) The zeros u_p of $a(u)$ are fixed points of T, i.e.
$u_p = Tu_p$ for $p = 1(1)\bar{n}$.

    c) The following monotonicity condition is satisfied:
$$v_\nu \to v \quad \text{for} \quad \nu \to \infty, \quad a(v) > 0$$
and
$$Tv_\nu \to w \quad \text{for} \quad \nu \to \infty,$$
implies
$$a(w) < a(v).$$

Then for any initial value $v_0 \epsilon B$, the iteration sequence $\{v_\nu\}$ defined by
$$v_{\nu+1} := Tv_\nu \quad \text{for} \quad \nu = 0, 1, \ldots$$

converges towards some fixed point u_p of T, i.e. there exists a natural number $p \in \{1(1)\bar{n}\}$ such that

$$\lim_{\nu \to \infty} v_\nu = u_p .$$

P r o o f : The space B being compact, we can extract a subsequence $\{v_{\nu(\mu)}\}$ which is convergent together with its image sequence, i.e. there exist elements $u^*, v^* \in B$ such that

$$v_{\nu(\mu)} \to u^* ,$$

$$Tv_{\nu(\mu)} \to v^* \quad \text{for} \quad \mu \to \infty$$

Hence,

$$a(v_{\nu(\mu)}) \to a(u^*) ,$$

$$a(Tv_{\nu(\mu)}) \to a(v^*) .$$

We want to show that $a(u^*) = 0$.

Let us assume $a(u^*) > 0$. By property c) we can conclude that

$$a(v^*) < a(u^*) .$$

Hence, for sufficiently large μ one obtains

$$a(Tv_{\nu(\mu)}) < a(u^*) .$$

This, however, contradicts the monotonicity relation

$$a(v_{\nu(\mu)}) > a(Tv_{\nu(\mu)}) > a(u^*),$$

with the result that u^* has to coincide with one of the zeros of $a(u)$, say u_{p_0}.

It remains to be shown that

$$v_\nu \to u_{p_0} \quad \text{for} \quad \nu \to \infty .$$

To this end we introduce the open sets

$$U(\delta) := \{u \in B \mid a(u) < \delta\}.$$

There exists a $\delta_0 > 0$ such that

$$U(\delta) = \bigcup_{p=1}^{\bar{n}} U_p(\delta) \quad \text{for} \quad 0 < \delta \leq \delta_0$$

where the $U_p(\delta)$ are open sets satisfying

$$U_p(\delta) \cap U_q(\delta) = \emptyset \quad \text{for} \quad p \neq q ,$$

$$U_p(\delta) \supset U_p(\delta-\varepsilon) \quad \text{for} \quad 0 < \varepsilon < \delta ,$$

$$u_p \in U_p(\delta) .$$

In addition for each $p \in \{1(1)\bar{n}\}$, the set

$$\{U_p(\delta_0/k) \mid k = 1,2,3,\dots\}$$

forms a neighborhood basis at the point u_p, i.e. in any given neighborhood of u_p at least one of the neighborhoods $U_p(\delta_0/k)$ is contained.

Let us now assume that our sequence $\{v_\nu := T^\nu v_0\}$ does not converge towards U_{p_0}.
Then for any neighborhood $U_{p_0}(\delta_0/k)$, $k = 1,2,3,\ldots$, there exists an index $\nu(k)$ such that $\nu(k) \to \infty$ for $k \to \infty$ and

$$v_{\nu(k)} \varepsilon\, U_{p_0}(\delta_0/k) \,,$$

$$Tv_{\nu(k)} \notin U_{p_0}(\delta_0/k) \,.$$

As necessarily $v_{\nu(k)} \neq u_{p_0}$, we have

$$a(Tv_{\nu(k)}) < a(v_{\nu(k)}) < \delta_0/k \,,$$

and, hence,

$$Tv_{\nu(k)} \varepsilon\, U_p(\delta_0/k) \quad \text{for some } p \neq p_0.$$

This, however, contradicts the continuity of T at u_{p_0}, and we obtain the sequence $\{v_\nu\}$ converging towards u_{p_0}. This completes the proof of the theorem.

6. Application of the abstract convergence theorem

The compact metric space of the above theorem will be a simply connected, bounded closed region B in the complex plane (with generic elements $u, v, w, z, \ldots$), large enough to contain all zeros w_p of the given polynomial $P(z)$, and such that the boundary

∂B of B is a level line of $|P(z)|$. Then $|P(z)|$ is larger for any z outside of B than for any z in B. Hence, the transformation T, as defined by the normal step and the emergency step of section 2, maps B into B.

Setting

$$a(z) = |P(z)|,$$

its continuity is obvious.

The continuity of T at the zeros w_p of $a(z)$ is easily verified. These zeros are obviously fixed points of T.

It remains to verify the monotonicity property c), i.e. the following

S t a t e m e n t :

Let be $v \varepsilon B$ with $P(v) \neq 0$. Let $\{v_\nu\}$ be a sequence with $v_\nu \varepsilon B$ for $\nu = 1, 2, \ldots$ and with $v_\nu \to v$. Let $Tv_\nu \to w$. Then the inequality $|P(w)| < |P(v)|$ holds.

P r o o f : Suppose there are infinitely many numbers v_ν such that the mapping Tv_ν consists in applying the normal step of Section 2 defined by (7). Then, because of the continuity of $P(z)$ and because of (6) one finds

$$|P(w)| = \lim_{\nu \to \infty} |P(Tv_\nu)| \leq (1-\varepsilon) \lim_{\nu \to \infty} |P(v_\nu)|$$

(15)

$$= (1-\varepsilon) |P(v)| < |P(v)|.$$

In this case the statement is true.

In what follows we may thus assume, without loss of generality, that application of the mapping T to each member of the sequence $\{v_\nu\}$ invariably means the emergency step of Section 2, defined by equations (8) to (11) (replacing the variable z^* by v_ν for fixed ν).

We will show then that in a certain neighborhood U of v the quantity

$$2^{-\overline{m}\ell-1} r^\ell |c_\ell/c_0|,$$

occurring in (10), is bounded away from zero, i.e. that there exists some $\gamma > 0$ such that

(16) $$2^{-\overline{m}\ell-1} r^\ell |c_\ell/c_0| \geq \gamma \quad \text{on} \quad U.$$

To this end we first observe that $c_j(z)$ and $r(z)$ are continuous at any point $z \epsilon B$. For $r(z)$ this results from the defining relation (5) if one takes into account that the functions

$$|c_0(z)/c_j(z)|^{1/j}, \quad j = 1(1)n$$

are each continuous at any given point z, except perhaps for some indices j for which $c_j(z) = 0$. These, however, are discarded in forming the minimum as required in (5).

Let now κ be such that

$$P'(v) = P''(v) = \ldots = P^{(\kappa-1)}(v) = 0, \; P^{(\kappa)}(v) \neq 0.$$

Then there exists a natural number M with

(17) $$|c_\kappa|\left(\frac{r}{2^M}\right)^\kappa \geq 3 \sum_{\substack{j=1 \\ j \neq \kappa}}^{n} |c_j|\left(\frac{r}{2^M}\right)^j \quad \text{at} \quad z = v.$$

As $c_j(z)$ and $r(z)$ are continuous functions of z, we deduce from (17) that in some neighborhood U of v we have

(18) $$|c_\kappa|\left(\frac{r}{2^M}\right)^\kappa \geq 2 \sum_{\substack{j=1 \\ j \neq \kappa}}^{n} |c_j|\left(\frac{r}{2^M}\right)^j.$$

With the same argument used to deduce (10) from (12) it is seen that on the neighborhood U the inequality (10) holds for $m = M$.

This shows that, on U, $\bar{m}$ is uniformly bounded by M.

It remains to evaluate γ (see (16)). Because of the definition of ℓ (see (9)) one has

$$2^{-\bar{m}\ell-1} r^\ell |c_\ell/c_0| \geq 2^{-Mn-1} r^n |c_n/c_0|.$$

By possibly restricting the neighborhood U, one can ensure that $c_0 := P(z)$ is bounded and that the continuous function $r(z)$ is bounded away from zero on U.

Then the following sequence of inequalities holds for all $v_\nu \varepsilon U$:

$$\begin{aligned}
|P(w)| &= \lim_{\nu \to \infty} |P(Tv_\nu)| \\
&\leq (1-\gamma) \lim_{\nu \to \infty} |P(v_\nu)| \\
&< \lim_{\nu \to \infty} |P(v_\nu)| = |P(v)|
\end{aligned}$$

which completes the proof of our statement. Therefore the algorithm defined and described in Section 2 is convergent.

Remark: This convergence proof is true only in the complete field of complex numbers. In a real computer, however, there is only a finite set of complex numbers, whence this convergence proof is meaningless in computer arithmetic. There still remains a certain "weak" convergence in the following sense: Because of the finiteness of the set of complex machine numbers and because of the monotonicity of the algorithm, there is a stop after a finite number of steps and the approximation $\hat{z}$ is "best" in the following sense: either $P(\hat{z}) = 0$ (under consideration of the round-off errors and the machine underflow) or the z increment is too small for the machine numbers.

Acknowledgment: The authors gratefully acknowledge stimulating criticism by Professor Peter Henrici.

TABLE I. Polynomial degree $n = 5(5)55, 100$. Roots distributed at random in the domain $|Re\ z| \leq 2$, $|Im\ z| \leq 2$. Program: POLROOT. Results with IBM 360/40 (double precision, 16 decimal digits).

Number of roots used in the investigation	Maximum number of halvings during emergency steps B	Mean number of halvings during emergency steps B	Maximum number of iterations	Mean number of iterations	Mean computation time in seconds for one root	Worst value of the relative error	Mean relative error	Polynomial degree n
100	34	2.70	20	6.55	1.9	$3.2_{10}-13$	$9.0_{10}-15$	5
200	44	2.77	17	8.10	3.2	$3.6_{10}-12$	$4.1_{10}-14$	10
300	37	3.96	19	8.90	4.9	$6.7_{10}-11$	$4.9_{10}-13$	15
400	44	4.35	21	9.30	7.3	$1.5_{10}-9$	$9.4_{10}-12$	20
500	47	4.61	20	9.56	10.0	$4.0_{10}-10$	$6.2_{10}-12$	25
600	113	5.71	25	10.03	12.7	$1.0_{10}-8$	$7.0_{10}-11$	30
350	58	5.09	18	9.98	13.6	$2.3_{10}-10$	$5.7_{10}-12$	35
320	30	3.94	21	9.97	15.6	$5.9_{10}-9$	$6.6_{10}-11$	40
450	88	5.94	20	10.44	18.1	$2.3_{10}-8$	$3.1_{10}-10$	45
500	68	6.00	22	10.40	21.7	$2.2_{10}-6$	$1.4_{10}-8$	50
440	72	6.00	20	10.49	25.3	$2.6_{10}-7$	$3.1_{10}-9$	55
300	97	8.31	19	11.37	64.2	$1.1_{10}-6$	$1.2_{10}-8$	100

Table II. Polynomial degree $n = 100$. Random valued polynomial roots w_p for $p = 1(1)100$. The roots w_p lie in the square $|Re\ z| \leq 2$, $|Im\ z| \leq 2$.

p	$Re\ w_p$	$Im\ w_p$
1	-8.523898124694824E-01	-1.132916688919067E+00
2	-1.064343452453612E+00	+4.300053119659422E-01
3	1.664720058441161E+00	-1.392654657363891E+00
4	3.324813842773436E-01	-1.422551870346068E+00
5	9.620118141174315E-01	-6.532597541809082E-02
6	8.023262023925781E-03	+3.456490039825438E-01
7	2.631182670593260E-01	+1.705514669418334E+00
8	1.264039993286132E+00	-1.205821752548217E+00
9	1.979222297668456E-01	-1.567437410354612E+00
10	-6.914606094360350E-01	+1.138856172561644E+00
11	-1.974434852600097E-01	+1.284763097763060E+00
12	-1.744220733642577E+00	+1.751753091812132E+00
13	1.519403934478758E+00	-9.126884937286376E-01
14	1.262674331665038E-01	-8.388106822967528E-01
15	1.970973014831542E-01	-6.783468723297118E-01
16	1.846761703491209E+00	+1.051735162734984E+00
17	9.051942825317382E-02	+1.125483274459838E+00
18	2.502698898315428E-01	+1.264210939407347E+00
19	-1.391968727111816E-01	+2.927472591400145E-01
20	1.076049804687499E+00	+7.956869602203368E-01
21	1.575331211090086E+00	-1.726359605789184E+00
22	-3.039073944091796E-02	+7.994835376739500E-01
23	-2.922635078430174E-01	-8.253920078277587E-01
24	-5.172939300537108E-01	-4.148290157318115E-01
25	-1.862098217010497E+00	-1.241654634475707E+00
26	-1.926651954650878E+00	+1.618569612503051E+00
27	-8.480401039123534E-01	+1.443797826766966E+00
28	-8.942070007324217E-01	-1.478250265121459E+00
29	4.422936439514160E-01	+1.306160688400267E+00
30	-1.494512557983398E-01	-1.926968336105345E+00
31	1.075358390808104E-01	-9.231207370758056E-01
32	4.843730926513670E-01	+1.206985712051390E+00
33	1.652569293975828E+00	-8.563511371612548E-01
34	1.090273857116699E+00	-9.855678081512449E-01
35	-1.011473178863525E+00	-1.449585199356078E+00
36	-1.342491149902342E+00	+1.742441415786742E+00

Table II (continued)

p	$Re\ w_p$	$Im\ w_p$
37	1.432791233062744E+00	-2.627611160278320E-03
38	8.158588409423828E-02	-2.056663036346435E-01
39	-1.291004657745360E+00	+8.409669399261473E-01
40	-8.357372283935545E-01	+7.296793460845947E-01
41	1.947021961212156E+00	-1.906106233596801E+00
42	1.613547325134276E+00	-1.735701322555540E+00
43	6.830039024353027E-01	+1.425098180770873E+00
44	-1.956302642822265E+00	+1.270262002944945E+00
45	-1.406760215759276E-01	-1.840224504470824E+00
46	9.752206802368163E-01	-1.224110364913939E+00
47	8.246150016784667E-01	-1.720628976821899E+00
48	1.834875106811523E+00	+9.657518863677977E-01
49	-1.666240215301513E+00	-5.784199237823484E-01
50	-4.433155059814452E-02	+1.806690692901611E-01
51	1.547891139984129E+00	-1.196520328521728E-01
52	-4.231414794921873E-01	-8.228850364685058E-02
53	-9.656081199645995E-01	-3.941299915313720E-01
54	-1.156046867370604E+00	-1.698005199432372E-01
55	-2.331051826477049E-01	+1.204591512680052E+00
56	-1.191289901733398E+00	+7.277033329010009E-01
57	-1.874717235565184E+00	-1.060792207717894E+00
58	4.291372299194334E-01	+1.576043367385864E+00
59	-1.104311466217040E+00	-2.713358402252196E-01
60	1.569492340087889E+00	+4.972898960113524E-01
61	-1.729505062103270E+00	+1.481559276580809E-01
62	-1.999716758728027E+00	-2.302367687225340E-01
63	8.483347892761230E-01	-5.708715915679931E-01
64	3.982677459716796E-01	-1.171966314315795E+00
65	6.340909004211425E-01	+4.592769145965576E-01
66	-6.535463333129881E-01	-7.370784282684324E-01
67	-1.961103916168212E+00	-1.217453241348265E+00
68	-1.665901184082030E+00	+1.821497201919555E+00
69	8.801331520080566E-01	-4.008667469024657E-01
70	-1.243288993835448E+00	-5.929191112518310E-01
71	1.381434917449950E+00	-1.234518289566040E+00
72	9.160480499267577E-01	-1.920757055282592E+00
73	1.172684192657470E+00	-2.057125568389891E-01
74	1.020117759704589E+00	-1.946196317672728E+00

Table II (continued)

p	$Re\ w_p$	$Im\ w_p$
75	2.900137901306152E-01	-1.145504236221312E+00
76	1.831779479980467E-01	-1.297166585922240E+00
77	-1.824193477630614E+00	+1.771301984786987E+00
78	1.425736427307128E+00	-4.453475475311278E-01
79	-1.321304798126219E+00	+1.026151418685913E+00
80	-1.325448989868163E+00	+1.293831110000609E+00
81	-9.464373588562011E-01	+7.567393779754638E-01
82	1.762629508972166E+00	-1.238810300827025E+00
83	-1.038458347320556E+00	+1.757011175155638E+00
84	1.429229736328124E+00	+1.953798532485961E+00
85	1.470015048980712E+00	-1.522837877273558E+00
86	-1.680140495300292E+00	+1.024977922439575E+00
87	-1.947384357452392E+00	+2.363753318786620E-02
88	-1.372337341308593E-02	-7.157018184661864E-01
89	1.589226245880126E+00	-8.408672809600828E-01
90	1.886488914489745E+00	-1.802420377731323E+00
91	-6.340203285217284E-01	+1.302592992782592E+00
92	3.847541809082030E-01	-1.613107442855833E+00
93	-1.924741268157958E+00	+1.529213666915892E+00
94	1.751580238342284E+00	+6.305572986602782E+01
95	8.156962394714355E-01	+1.570440053939819E+00
96	-8.362751007080076E-01	-1.136855840682982E+00
97	9.217500686645507E-02	-1.186032533645628E+00
98	1.704195976257323E+00	+1.175473451614379E+00
99	-1.184172153472899E+00	-6.962587833404540E-01
100	-6.377487182617187E-01	-1.185384511947630E+00

Table II (continued). The coefficients a_ν for $\nu = 0(1)100$.

ν	$Re\ a_\nu$	$Im\ a_\nu$
0	2.795304772536624E+15	+6.990662184765655E+14
1	1.158831097619413E+16	−2.756339144782078E+15
2	1.107221567777107E+17	−4.448776333218064E+16
3	2.901003846040368E+17	−4.255668491031971E+17
4	4.654403579570878E+17	−1.685078937843746E+18
5	−1.236877889009254E+18	−5.800187941146473E+18
6	−1.097757750359476E+19	−9.900989560071292E+18
7	−3.076155486041967E+19	−7.665572723859137E+18
8	−6.838509690553678E+19	+2.771769531101121E+19
9	−4.280975978484634E+19	+1.283832667198467E+20
10	9.067801143385135E+19	+1.604601100748268E+20
11	2.176057361859313E+20	+7.583054963301473E+19
12	2.518404157313454E+20	−1.270992039781286E+20
13	5.037826810996131E+19	−2.946042122905085E+20
14	−2.366294020849399E+20	−1.733404867560553E+20
15	−2.451085145817920E+20	+2.272481803055646E+20
16	2.320022304486213E+20	+4.352573562973662E+20
17	7.411495361196305E+20	+4.500032065403900E+19
18	7.359339237671072E+20	−7.478098910938799E+20
19	−1.785639338024548E+19	−1.363392847997088E+21
20	−1.129129021535609E+21	−1.187344986234427E+21
21	−1.848735039565895E+21	−1.570690755863954E+20
22	−1.593694706264291E+21	+1.159108154950287E+21
23	−4.857493011073617E+20	+1.922779406513329E+21
24	8.115791043734721E+20	+1.730081882927526E+21
25	1.579723557281554E+21	+7.663578093178774E+20
26	1.510103681055643E+21	−3,396512985739743E+20
27	8.188301992503817E+20	−1.022834048173106E+21
28	1.592656026536435E+18	−1.058843064678833E+21
29	−5.135539375722258E+20	−6.426203210589367E+20
30	−5.958485456463512E+20	−1.250216484416597E+20
31	−3.750191886933884E+20	+2.023006626998726E+20
32	−9.564017860013928E+19	+2.652874358358832E+20
33	7.347621535476306E+19	+1.562710261348914E+20
34	9.479912695028888E+19	+2.456402136771829E+19
35	3.764685299988986E+19	−4.117624077808775E+19
36	−2.083888798992704E+19	−3.430894423552400E+19
37	−3.782724006171410E+19	+4.016345833139755E+18

Table II (continued)

ν	$Re\ a_\nu$	$Im\ a_\nu$
38	-1.944826691806357E+19	+3.190376979973033E+19
39	9.192189351778953E+18	+3.278658165537994E+19
40	2.474243676554822E+19	+1.466172800677368E+19
41	2.264360556714000E+19	-4.703030213632054E+18
42	1.040713502459139E+19	-1.422887645778581E+19
43	-1.175806459487225E+18	-1.287239711165252E+19
44	-6.746116810961768E+18	-6.302770430714880E+18
45	-6.379173182836749E+18	-1.221360022814897E+17
46	-3.274998542663503E+18	+2.841873342642726E+18
47	-3.128226245607874E+17	+2.876558730479102E+18
48	1.135637039016613E+18	+1.511565697051102E+18
49	1.200798975943348E+18	+2.025141035497794E+17
50	6.334534547955278E+17	-4.311005729143565E+17
51	1.011807018927682E+18	-4.610859476986346E+17
52	-1.460412597275086E+17	-2.465970104946365E+17
53	-1.647870796236000E+17	-4.962102267782626E+16
54	-9.258666021002508E+16	+4.544516794019904E+16
55	-2.068221202629110E+16	+5.765574695560672E+16
56	1.572841104020418E+16	+3.206733235129135E+16
57	1.930520227840314E+16	+5.649286735465007E+15
58	9.297925654542235E+15	-5.760807075182194E+15
59	1.055858575655277E+15	-5.714234234172318E+15
60	-1.794471245447591E+15	-2.515458302301468E+15
61	-1.688763734284231E+15	-2.944356593826483E+14
62	-8.101774548999608E+14	+5.801379585955129E+14
63	-6.413063136986865E+13	+5.797460035919931E+14
64	2.297417072565851E+14	+2.354118060597877E+14
65	1.713529838701616E+14	-2.032068597135864E+13
66	4.269222289029132E+13	-7.097941813395224E+13
67	-1.187218577896938E+13	-3.541499767967711E+13
68	-1.504522788117603E+13	-7.747097690171950E+12
69	-8.649559914344531E+12	+2.013943211388361E+12
70	-2.860416949308672E+12	+4.591788027301129E+12
71	1.136698987479692E+12	+3.153115255827988E+12
72	1.847157961296244E+12	+4.753458560160001E+11
73	7.105272812388345E+11	-6.508463188768040E+11
74	-6.696050046213336E+10	-4.179558876397516E+11
75	-1.418983623083248E+11	-7.481148153699952E+10
76	-5.546665680713005E+10	+1.400883375484913E+10

Table II (continued)

ν	$Re\ a_\nu$	$Im\ a_\nu$
77	-1.944696007233029E+10	+1.991651445657766E+10
78	3.156302876772576E+08	+1.903428309313141E+10
79	1.125677745455890E+10	+6.585298243834019E+09
80	6.677937835289400E+09	-4.420385788443036E+09
81	-7.052795213257494E+08	-4.273189525216534E+09
82	-2.034420850472263E+09	-4.348772404835829E+08
83	-4.437651379261283E+08	+7.890245777395301E+08
84	2.851313308653548E+08	+2.515057250202284E+08
85	1.252644114944713E+08	-1.039256265236947E+08
86	-3.484274834264810E+07	-6.211928298270004E+07
87	-2.875486571644603E+07	+8.947256479871216E+06
88	1.179163398526755E+08	+1.202532842770203E+07
89	4.798525430588177E+06	+3.302464356788428E+05
90	5.213187315393187E+05	-1.916434662498561E+06
91	-7.144745662919332E+05	-4.329095474220228E+05
92	-2.587878750976153E+05	+1.953697204925875E+05
93	2.112951272680096E+04	+1.126715141826644E+05
94	3.395661188630288E+04	+1.105402891928841E+04
95	7.271783606137628E+03	-6.539967593655940E+03
96	-4.116011443173040E+02	-2.237979991400574E+03
97	-4.465371621069042E+02	-1.675160727199958E+02
98	-5.556594911040894E+01	+4.856734512468050E+01
99	8.867635726928710E-01	+1.044443321228027E+01
100	1.000000000000000E+00	+0.000000000000000E+00

Table II (continued). Results of the subroutine POLROOT. Computation with IBM 360/40 (double precision). Computing time 1 hour 47 minutes.

Current number in the sequence of evaluation	Subscript p of the evaluated root w_p	Number of iterations	Number of halvings during emergency steps B	Modulus of w_p: $\lvert w_p \rvert$	Relative error $\dfrac{\lvert \hat{z} - w_p \rvert}{\lvert w_p \rvert}$
1	38	10	4	2.21257E-01	3.68356E-15
2	88	15	40	7.15833E-01	3.61621E-12
3	15	10	9	7.06400E-01	3.44107E-12
4	50	10	6	1.86028E-01	2.06002E-14
5	52	11	4	4.31068E-01	2.24204E-15
6	14	13	31	8.48261E-01	7.93617E-11
7	31	12	17	9.29363E-01	1.57741E-10
8	6	7	5	3.45742E-01	4.06510E-15
9	23	13	7	8.75608E-01	7.70165E-13
10	19	9	1	3.24155E-01	1.43857E-14
11	75	17	32	1.18164E+00	4.56801E-10
12	24	7	0	6.63080E-01	6.20218E-14
13	63	10	11	1.02252E+00	3.32698E-12
14	69	11	7	9.67123E-01	5.48629E-13
15	40	13	16	1.10945E+00	2.96455E-13
16	53	16	12	1.04294E+00	1.31593E-13
17	5	10	2	9.64227E-01	1.05249E-14
18	64	15	28	1.23778E+00	4.80367E-10
19	22	10	1	8.00060E-01	6.46627E-14
20	59	14	9	1.13715E+00	1.87894E-13
21	76	13	25	1.31003E+00	3.70449E-10
22	10	13	13	1.33233E+00	4.90820E-14
23	34	11	12	1.46970E+00	9.27034E-10
24	17	11	11	1.12911E+00	7.79705E-13
25	66	10	5	9.85092E-01	8.30117E-14
26	73	11	7	1.19058E+00	9.55024E-14
27	81	11	9	1.21177E+00	6.55628E-13
28	65	9	2	7.82947E-01	3.90132E-15
29	4	18	31	1.46088E+00	2.48882E-09
30	2	12	5	1.14792E+00	6.70561E-14
31	46	10	8	1.56508E+00	3.55218E-09

Table II (continued)

Current number in the sequence of evaluation	Subscript p of the evaluated root w_p	number of iterations	Number of halvings during emergency steps B	Modulus of w_p: $\lvert w_p \rvert$	Relative error $\dfrac{\lvert \hat{z} - w_p \rvert}{\lvert w_p \rvert}$
32	55	15	6	1.22693E+00	1.50422E-13
33	89	11	24	1.79796E+00	1.80847E-08
34	54	12	6	1.16845E+00	8.47840E-14
35	97	9	9	1.18960E+00	2.48574E-10
36	39	14	13	1.54075E+00	1.07538E-11
37	13	11	16	1.77245E+00	3.71679E-08
38	32	13	7	1.30055E+00	9.36058E-13
39	18	9	1	1.28874E+00	1.19043E-12
40	96	16	20	1.41131E+00	2.58368E-12
41	78	11	6	1.49367E+00	9.03265E-12
42	56	15	5	1.39596E+00	3.85043E-12
43	100	14	7	1.34605E+00	5.02231E-13
44	37	9	12	1.43279E+00	9.67749E-14
45	91	11	7	1.44869E+00	4.51921E-14
46	8	11	9	1.74694E+00	1.46891E-07
47	79	15	10	1.67297E+00	2.37172E-11
48	47	16	38	1.90802E+00	1.02208E-08
49	29	8	5	1.37901E+00	1.34502E-12
50	99	13	6	1.37369E+00	5.41396E-14
51	70	9	4	1.37743E+00	6.72088E-14
52	9	10	5	1.57988E+00	4.77218E-10
53	43	9	3	1.58031E+00	4.96451E-14
54	80	10	8	1.85224E+00	2.34405E-11
55	71	13	10	1.85267E+00	4.21892E-07
56	60	8	1	1.64639E+00	8.13583E-16
57	27	12	7	1.67443E+00	8.08937E-13
58	1	8	8	1.41776E+00	1.99276E-12
59	85	13	18	2.11659E+00	4.87785E-07
60	11	12	4	1.29984E+00	2.35853E-13
61	3	15	12	2.17043E+00	2.39091E-07
62	20	10	3	1.33828E+00	1.87649E-14
63	61	12	14	1.73583E+-0	1.24955E-13
64	28	14	5	1.72766E+00	1.41007E-13
65	72	19	24	2.12801E+00	8.35193E-09

Table II (continued)

Current number in the sequence of evaluation	Subscript p of the evaluated root w_p	Number of iterations	Number of halvings during emergency steps B	Modulus of w_p: $\|w_p\|$	Relative error $\dfrac{\|\hat{z} - w_p\|}{\|w_p\|}$
66	26	14	27	2.51629E+00	2.69067E-09
67	51	9	0	1.55250E+00	3.24596E-13
68	12	16	18	2.47203E+00	5.65676E-09
69	49	10	6	1.76378E+00	7.16529E-14
70	42	16	14	2.36985E+00	9.65968E-07
71	94	14	4	1.86162E+00	6.93642E-15
72	58	10	6	1.63342E+00	1.61918E-14
73	87	18	24	1.94752E+00	6.28400E-14
74	92	11	6	1.65835E+00	1.75033E-09
75	36	10	8	2.19963E+00	3.73434E-11
76	82	10	6	2.15441E+00	4.41874E-08
77	86	14	6	1.96810E+00	1.19291E-11
78	16	13	5	2.12524E+00	3.26404E-14
79	57	13	9	2.15402E+00	7.91378E-13
80	90	13	12	2.60912E+00	3.78001E-08
81	7	9	3	1.72569E+00	3.32894E-14
82	35	11	3	1.76758E+00	1.04773E-13
83	77	11	7	2.54267E+00	3.90073E-09
84	33	8	3	1.86126E+00	1.07360E-08
85	48	13	12	2.07350E+00	3.55745E-14
86	62	10	2	2.01292E+00	1.03223E-13
87	45	17	19	1.84559E+00	8.54686E-12
88	98	11	1	2.07027E+00	7.29405E-15
89	30	9	2	1.93275E+00	4.61589E-12
90	68	11	6	2.46841E+00	1.28852E-09
91	95	10	3	1.76964E+00	5.27355E-15
92	21	12	4	2.33708E+00	1.05993E-06
93	44	12	6	2.33252E+00	5.57687E-11
94	25	13	2	2.23810E+00	1.30529E-12
95	83	9	2	2.04095E+00	1.18013E-12
96	74	8	3	2.19734E+00	5.18179E-09
97	41	8	0	2.72472E+00	1.07250E-08
98	93	9	2	2.45827E+00	1.72128E-09
99	84	4	2	2.42074E+00	4.67710E-16
100	67	0	0	2.30827E+00	1.41704E-12

TABLE III. Polynomial degree $n = 10$. Cluster of q roots in the domain $|Re\ z - 1| \le 0.01$, $|Im\ z| \le 0.01$; the other $n-q$ roots are distributed at random in the square $|Re\ z| \le 2$, $|Im\ z| \le 2$. Number of roots used for each q : 200. Program POLROOT. Results with IBM 360/40 (double precision, 16 decimal digits).

Maximum number of halvings during emergency steps B	Mean number of halvings during emergency steps B	Maximum number of iterations	Mean number of iterations	Worst value of the relative error	Mean relative error	
30	3.09	18	8.17	$2.1_{10}-12$	$3.4_{10}-14$	0
29	3.52	17	8.14	$8.8_{10}-13$	$2.2_{10}-14$	1
40	5.01	23	8.83	$7.5_{10}-12$	$3.4_{10}-13$	2
96	8.88	27	9.82	$3.9_{10}-9$	$1.0_{10}-10$	3
176	15.21	31	11.82	$5.3_{10}-6$	$7.4_{10}-8$	4
237	22.50	37	13.57	$8.7_{10}-5$	$3.1_{10}-6$	5
220	30.81	40	15.54	$3.0_{10}-3$	$3.5_{10}-4$	6
301	37.71	36	16.84	$1.5_{10}-2$	$4.6_{10}-3$	7
345	41.02	36	17.50	$2.7_{10}-2$	$1.2_{10}-2$	8
250	49.38	34	17.09	$3.7_{10}-2$	$2.2_{10}-2$	9
451	95.79	2	9.74	$5.1_{10}-2$	$2.5_{10}-2$	10

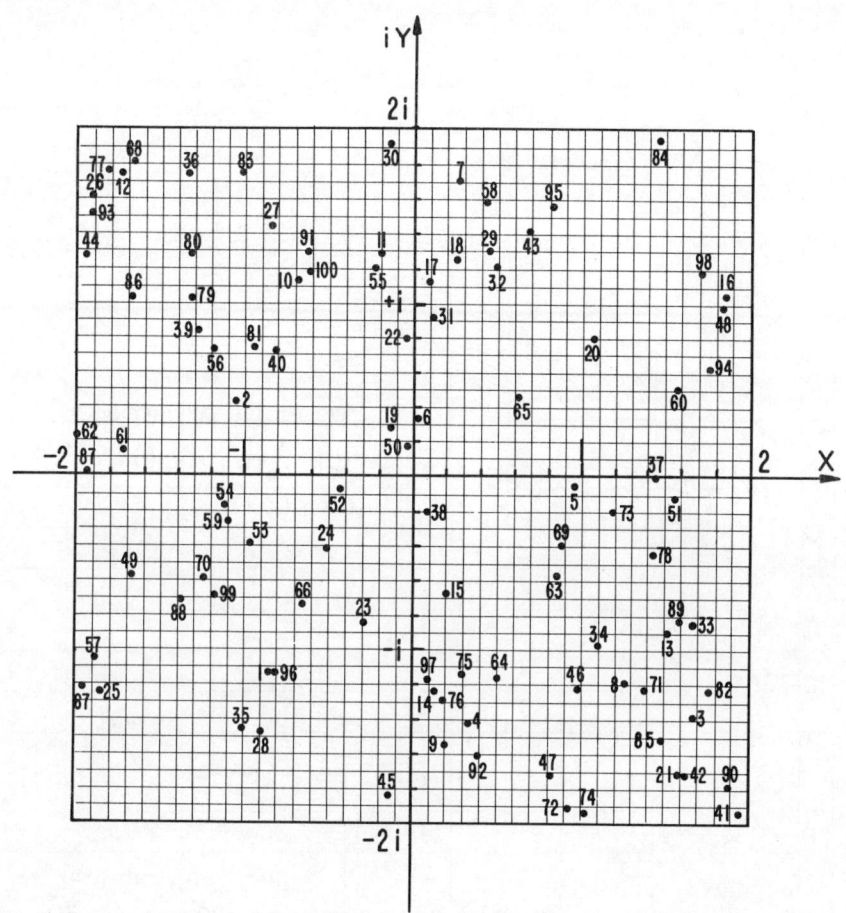

Fig. 1: Geometric distribution of the randomly generated roots w_p for $p=1(1)100$ in the square $|Re\ z| \leq 2, |Im\ z| \leq 2$.

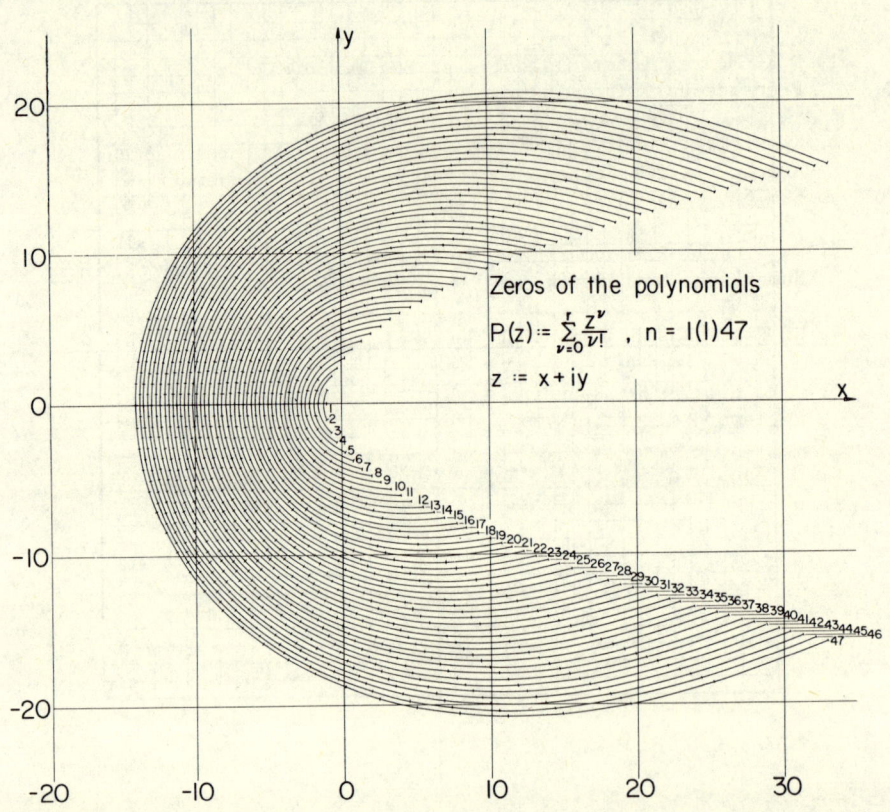

Fig. 2

REFERENCES

1. R.E. Moore: Interval analysis. Englewood Cliffs, (N.J.), Prentice-Hall Inc. 1966.

2. K. Nickel: Ueber die Notwendigkeit einer Fehlerschranken-Arithmetik für Rechenautomaten. Numerische Mathematik *9* (1966), 69-76.

3. K. Nickel: Die numerische Berechnung der Wurzeln eines Polynoms. Numerische Mathematik *9* (1966), 80-98.

4. H. Rutishauser: Computing *1* (1966), 161.

5. J.H. Wilkinson: Rounding errors in algebraic processes. Englewood Cliffs, (N.J.), Prentice-Hall Inc. 1963.

6. N. Apostolatos et al.: The Algorithmic Language Triplex-ALGOL 60. Numerische Mathematik *11* (1968), 175-180.

Prof. K. Nickel
Technische Hochschule
D-75 Karlsruhe
Germany

and

Dr. B. Dejon
IBM Zurich Research Laboratory
CH-8803 Rüschlikon
Switzerland

T. J. Dekker

Finding a Zero by Means of Successive Linear Interpolation

1. Introduction

In this contribution a procedure is described which searches for a zero of a given function in a given interval. The method is a mixture of linear interpolation and extrapolation, and bisection. Convergence to a zero, more precisely: an interval which is smaller than a given tolerance and in which the function changes sign, is guaranteed, if the function has different sign at the endpoints of the given interval. Moreover, the procedure has a completely satisfactory asymptotic behavior.

The procedure has been used extensively at the Mathematical Centre, Amsterdam, in two ALGOL 60 versions ([1], AP 200 and AP 230, a slightly improved version of which is published below) and at N.P.L., Teddington, England (see [2]).

2. The basic formula

Suppose we are given a real function f of one real variable. The value of f for any real x is denoted by fx.

Let a and b be two argument values of f. The linear function L, coinciding with f at a and b, is given by the formula

(2.1) $$L(x) = \frac{x \times (fb - fa) - (a \times fb - b \times fa)}{b - a}.$$

Choosing i such that $L(i) = 0$, we obtain the linear interpolation formula (also known as "regula falsi")

$$(2.2) \qquad i = \frac{a \times fb - b \times fa}{fb - fa}.$$

In the sequel "interpolation" will imply that fa and fb have different sign; otherwise, the term "extrapolation" will be used.

If the second derivative of f exists in an interval J including a, b and x, there exists a $\xi \in J$ such that

$$(2.3) \qquad f(x) = L(x) + \frac{1}{2}(x-a)(x-b)f''(\xi)$$

(cf. [4] formula (1.5) and (3.3)).

In particular, for a zero, z, of f, we obtain

$$(2.4) \qquad L(z) = -\frac{1}{2}(a-z)(b-z)f''(\xi).$$

On the other hand, since L is a linear function and $L(i) = 0$, we have

$$(2.5) \qquad L(z) = (z-i)\frac{fb - fa}{b - a} = (z-i)f'(\eta),$$

where η is between a and b. Combining (2.4) and (2.5) we find

$$(2.6) \qquad i-z = (a-z)(b-z)\frac{f''(\xi)}{2f'(\eta)},$$

where ξ lies in the smallest interval containing a, b and z, and η is between a and b.

3. Successive linear interpolation or extrapolation

We consider the iteration process which, starting from two given argument values a and b, successively applies formula (2.2), where in each step a and b are the two most recent iterates and i is the next iterate. This process does not always converge (the iterate i may even become infinite), but, if it does converge to a simple zero z of a function f having a continuous second derivative, the convergence is superlinear.

Indeed, since z is a simple zero, i.e. $f'(z) \neq 0$, and f'' and f' are continuous, the factor $\frac{1}{2}f''(\xi)/f'(\eta)$ occuring in (2.6), converges to $\frac{1}{2}f''(z)/f'(z)$ and is therefore bounded in a neighbourhood of z by some constant, K. So, if a and b are in this neighbourhood of z, we have

$$|i-z| \leq K|(a-z)(b-z)| \ .$$

Hence the ratio $|i-z|/|b-z|$ of two successive errors converges to zero because the error $|a-z|$ does; in other words, the convergence is superlinear. It can be proved that the order m of the process, i.e. the asymptotic value of $ln|i-z|/ln|b-z|$, equals the largest root of the equation

$$m^2 - m - 1 = 0 \ ,$$

thus

$$m = \frac{1}{2}(1 + \sqrt{5}) \approx 1.618.$$

(For more details see [4] and [2]).

4. Use of interpolations only

The danger of divergence can be avoided simply by performing only interpolations. Starting with two given argument values a and b satisfying $sign(fa) \neq sign(fb)$ (note that $sign(x)$ equals $+1$ for $x > 0$, 0 for $x = 0$, -1 for $x < 0$, cf. [3] 3.2.4), then in each step the next iterate i is obtained from the interpolation formula (2.2), where now b is the last iterate and a the most recent iterate satisfying $sign(fa) \neq sign(fb)$.

This process always converges, and, if f is continuous in the given interval, the limit is a zero of f. In many cases, however, the convergence is merely linear, because a may remain constant.

Indeed, assuming that f'' is continuous and f' does not vanish in the given interval, formula (2.6) yields

$$|i-z|/|b-z| \leq constant \times |a-z|,$$

where the constant is an upper bound of $\frac{1}{2}|f''(\xi)/f'(\eta)|$ in the given interval. So, if a remains fixed, the convergence is linear.

5. Guaranteed interval reduction with superlinear convergence

We modify the process described in Section 3 as follows. At each step linear interpolation or extrapolation is performed between the two most recent iterates. If, however, extrapolation yields a point outside the smallest known interval in which the function changes sign, this point is rejected and replaced by the midpoint of the interval.

In more detail, at each step three points a, b and c are involved. The point b is the most recent iterate, a the previous one, and the "contrapoint" c is the last iterate satisfying $sign(fb) \ne sign(fc)$. Interpolation or extrapolation is always performed between a and b, yielding according to formula (2.2), a provisional value i for the next iterate. The process starts with two points c and b, satisfying the above conditions, and with $a = c$. Thus the first step is an interpolation step yielding a point i between c and b.

To prepare the next step, the assignments "$a:=b$; $fa:=fb$; $b:=i$; $fb:=fi$" are carried out.

We now have two possibilities for the new function value fb.

5.1) $sign(fb) \ne sign(fc)$. Then the new values a, b and c obviously satisfy the above conditions. In the next step, extrapolation takes place between the most recent iterates a and b. If the value i obtained is not between c and b, it is replaced by the midpoint $m = (c+b)/2$.

5.2) $sign(fb) = sign(fc)$. Then obviously $sign(fb) \ne sign(fa)$, or, in other words, a is the new contrapoint. After performing the assignments "$c:=a$; $fc:=fa$", the process is ready for the next step, which is an interpolation step like the first one.

The asymptotic behavior of this process is the same as that described in Section 3 because, sooner or later, the iterates are so near the limit that no further rejections of extrapolated values occur.

Nevertheless, this process is not completely satisfactory for two reasons.

5.3) Firstly, the distance $|c-b|$ does not always converge to zero. This happens if two successive iterates lie on one side of the zero and the function is convex (i.e. $f(x)f''(x) > 0$). E.g. for the function $f(x) = x^3 + x$, any non-trivial pair of starting values c,b leads to this situation.

5.4) Secondly, convergence to a zero is guaranteed only for exact arithmetic. In a numerical process, however, two successive iterates may coincide, due to rounding errors, so that no convergence to a zero occurs. E.g. for the function (cf. [2])

$$f(x) = x(x-1)^5,$$

linear interpolation, performed between $a = -.5$ and $b = +.99$, would yield a value i so close to b, that, on a ten-decimal digit computer, i might be numerically equal to b. Also if the successive iterates do not coincide, convergence may be very slow in a case like this.

6. The final modifications

For these reasons, the process described above is modified in the following two respects.

6.1) Firstly, in order to have the problem of nearly coinciding iterates at only one endpoint, b, of the interval, the points b and c are interchanged whenever $|fc| < |fb|$.

If the iterates are near the limit, usually $|fb| \ll |fc|$, so the interchanges seldom occur and do not influence the asymptotic behavior of the process. The interchanges are carried out such that the new value of a will be equal to the new value of c, and so, after interchanging, the next step is always an interpolation step. In the situation (5.1) this means that extrapolation, which would not yield an acceptable value in this case, is discarded in favor of interpolation.

Because of the interchanges, $|fb| \leq |fc|$ now holds in each step, so that the next iterate should lie between b and $m = (c+b)/2$, which criterion is used as acceptance criterion (see below at 6.2).

6.2) Secondly, the difference of any two iterates is forced to be not smaller than some tolerance "tol". This is reached in each step as follows.

If $|b-i| \leq tol$, then i is replaced by $s = sign(c-b) \times tol+b$. Otherwise, if i is not between b and m, i is replaced by m.

Thus, the accepted value i is always between c and b, and, as long as $|c-b| \geq 2 \times tol$, the distance between i and the endpoints b and c is never smaller than tol.

In this way, both difficulties, mentioned in Section 5, are avoided. This is obvious for the second difficulty (5.4). As to (5.3), if the successive iterates remain on one side of a zero, then, at a certain stage, i will have to be replaced by s, which is beyond the zero in many cases (otherwise, it will certainly happen after a few more steps), so that the points c and b will satisfy $|c-b| \approx tol$ in the following step.

6.3) Remark

In the ALGOL 60 procedure, published below (contrary to the versions in [1]), the acceptance test for each new iterate is carried out before the division occurring in (2.2), in order to prevent overflow. In more detail: firstly the numerator and the denominator of (2.2) are calculated, these values then being used for checking whether the new iterate has to be equal to either s or m, and, only if this is not the case, the division is carried out.

7. The stopping criterion

The stopping criterion used is $|m-b| \leq tol$; i.e. $|c-b| \leq 2 \times tol$. In each step, the distance $|c-b|$ is diminished by at least the tolerance, so that after a finite number of steps, the stopping criterion is satisfied. The factor 2 is necessary, because otherwise, after finding different signs of the function at b and s, an extra iteration would be needed, if, due to rounding errors, $|s-b|$ were slightly greater than tol. Moreover, the factor 2 ensures that by replacing i by s the iterate does not get too close to c.

In the procedure published below, the given tolerance is a function of b, which, of course, must never be smaller than the machine precision. A suitable tolerance function, which is used in the versions published in [1], is $|b| \times re + ae$, where re is the desired relative precision and ae is an absolute tolerance, which must be chosen unequal to 0, if the given interval contains the origin.

8. Searching for a zero

The procedure published below may (in contrast with the versions in [1]) also be used to search for a zero, if no points having function values with different signs are known. If the given function has different signs at the two given points, the procedure determines a zero in the way described above.

If, on the other hand, the given function has the same sign at the two given points, the interval is reduced by means of either bisection, at each step choosing the half interval which has the smallest absolute value of the function at the outer endpoint, or by taking s as new iterate (see 6.2), if the function nearly vanishes at one endpoint. If sign change is detected, the procedure continues as above, otherwise the process ends on account of the above stopping criterion (see Section 7).

9. The ALGOL 60 procedure

The required data and the results of the procedure are described in the following comment, which preceeds the text of the procedure.

comment *zeroin* searches for a zero of a function between the given values of x and y within a certain tolerance. The function and the tolerance are, in this order, given by the actual parameters for fx and $tolx$, which are expressions depending on the Jensen variable x.

zeroin:= **true**, if either the function values at the given points x and y have different signs, or the procedure finds some point in between at which the sign of the function value differs from that at x and y. Then *zeroin* calculates and delivers two values x and y lying within the given interval, having function values of different signs and satisfying $abs(x-y) \leq 2 \times tolx$.

Moreover, the absolute function value is not greater at x than at y, so that the delivered value of x is the best value for the zero.

If the function has a continuous second derivative, the order of convergence is about 1.6.

$zeroin$:= **false** , if the procedure fails to find points at which the function values have different signs. Then the delivered values of x and y satisfy all the above conditions, except the sign change condition.

One has to take care that $tolx$ is never smaller than the machine precision. Then in either case the process is completed after a finite number of steps, an upper bound for the required number of steps being the length of the given interval divided by the minimum of the tolerance;

```
boolean procedure    zeroin(x, y, fx, tolx); real    x, y, fx, tolx;
begin real    a, fa, b, fb, c, fc, tol, m, p, q;
    a:= x; fa:= fx; b:= x:= y; fb:= fx;
interpolate: c:= a; fc:= fa;
extrapolate: if  abs(fc) < abs(fb) then
    begin   a:= b; fa:= fb; x:= b:= c; fb:= fc; c:= a; fc:= fa
    end interchange;
    tol:= tolx; m:= (c + b) × .5; if  abs(m - b) > tol then
    begin   p:= (b - a) × fb; if   p ≥ 0 then   q:= fa - fb else
        begin   q:= fb - fa; p:= - p end ;
        a:= b; fa:= fb;
        x:= b:= if   p ≤ abs(q) × tol then  sign(c - b) × tol + b else
        if  p < (m - b) × q then  p / q + b else   m; fb:= fx;
        goto if   sign(fb) = sign(fc) then   interpolate else
        extrapolate
    end ;
    y:= c; zeroin:= sign(fb) × sign(fc) ≤ 0
end zeroin;
```

Acknowledgement

The idea of using linear interpolation and extrapolation, rejecting the latter only if the value is unacceptable (see Section 5) is due to Prof. Dr. Ir. A. van Wijngaarden. Discussions and numerical experiments, in which Prof. Dr. E.W. Dijkstra, Dr. J.A. Zonneveld and the author also participated, led to the modification described in Section 6 and to the procedures published in [1] (AP 200 and AP 230). The idea of avoiding unnecessary divisions with possible overflow (see remark 6.3) is due to B.J. Mailloux M.Sc.

REFERENCES

1. T.J. Dekker (ed.), The Series AP 200 of Procedures in ALGOL 60 (Mathematical Centre, Amsterdam, 1962-1965).

2. J.H. Wilkinson, Two algorithms based on successive linear interpolation (Stanford University, Techn. Rep. no. CS 60, 1967).

3. P. Naur (ed.), Revised Report on the algorithmic language ALGOL 60 (1962).

4. A.M. Ostrowski, Solution of equations and systems of equations (2nd edition, New York 1966).

Dr. T.J. Dekker
Mathematisch Centrum
2e Boerhaavestraat 49
Amsterdam (O)

George E. Forsythe

Remarks on the Paper by Dekker

In the quantized world of floating-point digital computation with prescribed precision, "continuity" and "smoothness" of a function f cannot have their mathematical meanings. For $f(x)$ is computed only on a discrete point set S, and for even the smoothest of mathematical functions the computed values $f(x)$ are normally infected with a wide band of noise. For example, the computed values of a polynomial of low degree may have hundreds of sign changes in the neighborhood of a simple mathematical zero, and there may be no x in S for which the computed value of $f(x)$ is 0.

In digital computing continuity, smoothness, order of convergence, etc. become statistical concepts with meaning only in relation to intervals of x that are much larger than the quantum steps between adjacent points of S.

The rootfinder *zeroin* presented by Dr. Dekker has the following important advantages:

(a) It makes no assumptions about the smoothness of f, or even its continuity. *Zeroin* merely localizes an interval of sign-change of the computed values $f(x)$. Thus the algorithm is entirely safe at the microscopic level of quantized digital computing.

(b) For the same reason it can deal with functions f that are not even continuous in the mathematical sense. (For such functions the method is at least as good as bisection.) Similarly, it can deal efficiently with smooth functions f whose behavior in the large is quite different from that in the neighborhood of the zero. See [1].

(c) For functions that are smooth, *zeroin* enjoys the fast "convergence" properties of linear interpolation. That is to say, there is rapid progress of x from a value within an appropriate neighborhood of the desired zero z to values of x near (but not too near) z.

(d) Because the sought-for zero (i.e., sign change) is always bounded in an interval, there is no danger of unknowingly stopping the iteration while the iterate is still far from the zero. This danger is usually present in algorithms like simple linear interpolation, particularly for multiple zeros.

The algorithm is thus safe and effective. It is also rather subtle, and merits careful study.

It is unfortunate that *zeroin* has not yet been published, although it has been privately available for some years. J.H. Wilkinson [1] has discussed the method, along with a related algorithm of D.J. Wheeler [2]. For my knowledge of *zeroin*, I am indebted to Dr. Wilkinson, who recommends it highly.

REFERENCES

1. J.H. Wilkinson: "Two algorithms based on successive linear interpolation", Tech.Rep. CS 60, Computer Science Dept., Stanford University, Stanford, California, USA, April 1967.

2. M.V. Wilkes, D.J. Wheeler, and S. Gill: "The Preparation of Programs for a Digital Computer", Addison-Wesley, Reading, Massachusetts, USA, 1951.

Prof. G.E. Forsythe
Stanford University
Stanford, California 94305
USA

George E. Forsythe[*]

What is a Satisfactory Quadratic Equation Solver?

This Symposium has dealt with provable algorithms for finding zeros of general polynomials, with the tacit assumption that the processes would be implemented on an ideal computer system capable of exact arithmetic operations. In contrast, I should like to point out the near absence of algorithms to solve even a quadratic equation in a satisfactory way on actually used digital computer systems. The difficulties are partly caused by round-off error in floating-point arithmetic, but much more by the ever-present possibility of overflow or underflow (defined below).

This note presents specifications for a satisfactory quadratic equation solver suggested by Professor W. Kahan of the University of Toronto in lectures at Stanford University in 1966. The general level of performance is Kahan's, but the details are mine.

Consider the following set $F = F(\beta, t, m, M)$ of normalized floating-point numbers. This uses a number base β (bases $2, 8, 10$, and 16 are in use) and a prescribed number t of significant digits.

---

[*] This is a brief written contribution to Symposium on Constructive Aspects of the Fundamental Theorem of Algebra, held at the IBM Research Laboratories, Rüschlikon, Switzerland, 5-7 June 1967. The preparation of this note and the author's attendance at the Symposium were sponsored by the U.S. Office of Naval Research under project NR 044 211.

There are two limiting integer exponents m and M. The set F contains precisely

$$(1) \qquad 1 + 2(\beta - 1)(M - m + 1)\beta^{t-1}$$

numbers. One of these is 0. Each other number in F has a unique representation

$$(2) \qquad \pm N \times \beta^e ,$$

where the *sign* is + or −, where the integer *exponent* e satisfies the inequality

$$(3) \qquad m \leq e \leq M,$$

and where the *integer significand* N satisfies the normalization condition

$$(4) \qquad \beta^{t-1} \leq N \leq \beta^t - 1.$$

Frequently 0 is given the unique computer representation

$$(5) \qquad + 0 \times \beta^m .$$

See [1] for a discussion of this floating-point number system in a slightly different notation.

We choose F^* to be a certain subset of F consisting of numbers not too close to overflow or underflow.

DEFINITION. To be definite (and somewhat arbitrary), let $F^* = F^*(\beta,t,m,M)$ be the set consisting of 0 and all numbers (2) subject to (4) and also to

(3^*) $\qquad m + 1 \leq e \leq M - 1.$

DEFINITIONS. A real number x is said to be *in the range of* F^* if either $x = 0$ or

(6) $\qquad \beta^{t-1}\beta^{m+1} \leq |x| \leq (\beta^t - 1)\beta^{M-1}.$

A complex number z is *in the range of* F^* if both $Re(z)$ and $Im(z)$ are real numbers in the range of F^*.

One similarly defines the expression *in the range of* F for real and complex numbers.

The statement (in computer jargon) that a real number y suffers from *over-* or *underflow* is equivalent to saying that y is not in the range of F.

In terms of these concepts I now give specifications in the form of a commented heading in Algol 60 of what I consider to be a satisfactory quadratic-equation solver for use with a processor of floating-point numbers in $F(\beta,t,m,M)$.

procedure *QUADRATIC (a, b, c, x1, y1, x2, y2, error)*;
   value   *a, b, c*; **real** *a, b, c, x1, y1, x2, y2*; **switch** *error*;
comment We are solving the equation $az^2 + bz + c = 0$, for arbitrary input parameters a, b, c in F^*. Where values of the output parameters are not specified, they are irrelevant.

If $a = b = c = 0$, exit to $error[1]$, since all complex numbers z satisfy the equation.

If $a = b = 0$ and $c \neq 0$, exit to $error[2]$, since no z satisfies the equation.

Otherwise, let z_1 and z_2 be the exact roots of the equation, numbered so that $|z_1| \leq |z_2|$. (If $a = 0$, let $z_2 = \infty$.)

Whenever z_1 is in the range of F^*, set $x1$ and $y1$ to numbers in F close (defined below) to the real resp. imaginary parts of z_1.

Whenever z_2 is in the range of F^*, set $x2$ and $y2$ to numbers in F close to the real resp. imaginary parts of z_2.

Let $\zeta_1 = x1 + i \times (y1)$. We require that $\zeta_1 = 0$ (if $z_1 = 0$), and otherwise (again being somewhat arbitrary) that $|\zeta_1 - z_1| \leq \beta + 1$ units in the least-significant digit of the significand of the floating-point representation (2) of $max(|x1|, |y1|)$. To repeat this requirement in symbols, if $entier[x]$ denotes the greatest integer $\leq x$, we demand that

(7) $$|\zeta_1 - z_1| \leq (\beta + 1) \times \beta^e,$$

where $e = entier[log_\beta max(|x1|, |y1|) - t]$.
We make a corresponding requirement of $\zeta_2 = x2 + i \times (y2)$.

If z_1 is not in the range of F, but z_2 is in the range of F^*, set $x2$, $y2$ as above and exit to $error[3]$.

If z_2 is not in the range of F (including the case $z_2 = \infty$), but z_1 is in the range of F^* set $x1$, $y1$ as above and exit to $error[4]$.

If neither z_1 nor z_2 is in the range of F, exit to $error[5]$.

If a root z_i is in the range of F but not in the range of F^*, we permit either an indication of over- or underflow via the appropriate exit to *error*, or a determination of z_i with an accuracy satisfying (7) above.

The procedure QUADRATIC should make no unnecessary use of multiple-precision computation, but computation with $2t$ significant digits is essential at one part of the procedure, to achieve the accuracy (7). End of comment;

The main source of practical difficulty in writing the procedure QUADRATIC is the possibility of over- or underflow in many places. What is actually programmed depends crucially on what the computing system does in case of over- or underflow. An ideal system permits the user's program to regain control of the algorithm, if the user wishes, with the ability to interrogate Boolean variables to learn whether there has been overflow, underflow, or neither. Though such systems are rare, they have been implemented at Toronto [3] and at Stanford [2], and these systems make programming such an algorithm as QUADRATIC far more satisfactory.

Some systems merely dismiss a user's program in case of overflow. If a result underflows, many systems merely set the result to 0 and return to the user's program without any indication. Faced with systems like these, the programmer must take great pains to insure that over- or underflow can never occur. These precautions make a satisfactory algorithm tedious to write, lengthy to store, and slow to execute. One necessary subroutine must determine the exponents of a, b, c, and other real numbers local to the procedure. This is probably most gracefully written in machine code.

The ability to achieve the prescribed accuracy (7) of $\beta + 1$ units in the last place depends on the detailed properties of the floating-point arithmetic processor. If the prescribed accuracy is not achievable, condition (7) must be relaxed as necessary. If necessary, one could also make the set F^* smaller by changing (3^*).

As a simple illustration suppose that $\beta = 10$, $t = 4$, $m = -54$, $M = 45$. Then the smallest and largest positive numbers in F are

$$1000 \times 10^{-54} = 10^{-51}$$

and

$$9999 \times 10^{45} = .9999 \times 10^{49} = (1 - 10^{-4}) \times 10^{49}.$$

Here are some equations that may give trouble for this system:

(a) $\quad 10^{-40} z^2 - 5 \times 10^{-40} z + 6 \times 10^{-40} = 0$:

The roots are 2 and 3, and the only danger is that undetected underflows will introduce an error.

(b) $\quad z^2 + 10^{10} z - 1 = 0$:

Use of the standard quadratic formula will yield 0 for the positive root, instead of the correctly rounded value 1.000×10^{-10}.

(c) $\quad 10^{-30} z^2 - 10^{40} + 1 = 0$:

The roots are near 10^{70} and 10^{-40}; use of the quadratic formula can easily cause overflow or underflow.

(d) $2.864\ z^2 - 2.864\ z + 0.7160 = 0$:

Here 0.5000 is a double root. Evaluating the quadratic formula

$$\frac{2.864 \pm \sqrt{(2.864)^2 - (4 \times 2.864) \times 0.7160}}{2 \times 2.864}$$

in single-precision rounded arithmetic yields $0.5000 \pm 0.05477\ i$, with an error of over 547 units in the last place of 0.5000.

It is not purely academic to make strict demands of the procedure QUADRATIC. Quadratic equations arising in the course of solving determinantal equations by Muller's method [6] or Laguerre's method [5], particularly in connection with large matrices, sometimes have one of the roots out of the range of F^*, and yet make essential use of the root in the range of F^*. The accuracy requirement (7) is perhaps overstrict for equations with nearly double roots.

I believe the specifications to be very reasonable for a basic process like solving a quadratic equation. Nevertheless, I venture to guess that not more than five quadratic solvers exist anywhere that meet the general level of the specifications. Kahan [4] has prepared an algorithm (in Fortran IV for the 7094-II under the Toronto version of the IBSYS operating system) which achieves the specifications for $\beta = 2$, $t = 27$, $m = -155$, $M = 100$. The error never exceeds $9/4$ ($= \beta + 1/4$) units in the least-significant place of the root. The only multiple-precision operations occur in the computation of $\Delta = b_1^2 - 4a_1c_1$, followed by storage of a single-precision value of Δ. (Here a_1, b_1, c_1 are proportional to a, b, c.)

The excess time for the double-precision computation is negligible in comparison with the time required to deal with over- and underflow.

It is obviously relevant to ask to what extent the various computer algorithms presented at this Symposium for general polynomials make provision for over- and underflow, and also what accuracy they achieve.

It is noteworthy that the programming of *QUADRATIC* depends crucially on the arithmetic properties of the computing system, especially on its behavior with over- and underflow. The practical numerical analyst with high standards is thus inextricably involved with the arithmetic behavior of his digital computer hardware and accompanying operating systems. Unfortunately, few numerical analysts have formulated their systems requirements explicitly, not to mention communicating them effectively to the persons who design hardware and software systems. With existing computing systems a numerical analyst faces a most disagreeable dilemma -- either he writes less than satisfactory algorithms, or he undertakes the never-ending chore of writing basic software systems (and perhaps even rebuilds the arithmetic unit). Professor Kahan has generally taken the second alternative.

REFERENCES

1. George E. Forsythe and Cleve B. Moler: "Computer Solution of Linear Algebraic Systems", Prentice-Hall, Englewood Cliffs, N.J., 1967; see Sec. 20.

2. Michael D. Green: "Coping with over/underflow on the B5500", multilithed typescript, Computer Science Dept., Stanford University, Stanford Calif. 94305, USA, 15 July 1966, c. 50 pp.

3. W. Kahan: "7094-II system support for numerical analysis", reproduced by and available from SHARE as item C-4537 in SSD 159 (Dec. 1966).

4. W. Kahan: "The FORTRAN IV subroutine QDRTC", computer library of McLennan Laboratories, University of Toronto, Toronto, Canada, c. 1966.

5. H.J. Maehly: "Zur iterativen Auflösung algebraischer Gleichungen", Z. Angew. Math. Phys., 5 (1954) 260-263.

6. David E. Muller: "A method for solving algebraic equations using an automatic computer", Math. Tables Other Aids Comput., 10 (1956) 208-215.

Prof. G.E. Forsythe
Stanford University
Stanford, California 94305
USA

L. Fox

Mathematical and Physical Polynomials

There are really two different problems associated with polynomials, and both the input and the output of a computer program for determining zeros should be different for these two problems. The first problem is what might be called *mathematical*, meaning that the data, the coefficients of the polynomial, are known exactly. The author of such a problem has the right to ask for solutions of unlimited accuracy, and in general the ideal program will contain a parameter, to be set in advance, specifying the required number of accurate digits in the computed solution.

Now it may not be possible to *store* the data exactly, even if its mathematical form is exact. For example, numbers like π, $sin\ 0.7$, or even the exact 0.6 (decimal), cannot be stored exactly in a binary machine. In single-precision (floating-point) arithmetic, therefore, instead of starting with the given polynomial equation $\sum_{k=0}^{n} a_k z^k = 0$, we actually use the slightly "perturbed" problem

(1) $$\sum_{k=0}^{n} (a_k + \delta_M a_k)\ z^k = 0.$$

Here the subscript M means "mathematical", and $\delta_M a_k$ is the *mathematical error*, of rounding or truncation to a finite number of digits. In fact if t digits are used in floating-binary arithmetic for the fractional part of a_k, then

$$|\delta_M a_k| \le 2^{-t}|a_k| . \tag{2}$$

There is, however, yet another source of error. In determining a zero of the polynomial a fundamental operation is the computation of the polynomial for a given argument z, which we can certainly *choose* to have exact digital storage in the computer. By the process of backward error analysis (Wilkinson, 1963) we can show, for example in the technique of "nested multiplication", that the error in the computation can be "thrown back" into the data, so that we are again using a further perturbed polynomial. This extra perturbation, denoted by $\delta_T a_k$, where T means "technique", depends in value on the particular argument z, but its upper bound is independent of z. More important is the fact that the perturbation is a function of the method or technique.

The result of all this is that the problem we actually solve is a perturbation

$$\sum_{k=0}^{n} (a_k + \delta_M a_k + \delta_T a_k) z^k = 0 \tag{3}$$

of the original problem. Two important consequences follow. First, the size of the δ_T perturbation gives a relative evaluation of numerical techniques. Those for which $\delta_T a_k$ is large exhibit what might be called "induced instability", the adjective denoting that the instability is due to the method and not to the nature of the problem. In particular we shall fail to determine accurately even "well-conditioned" zeros if $\delta_T a_k$ is large, and a particular instance

of this is the process of deflation, mentioned in Dr. Traub's
paper, in which the zeros are removed in decreasing order of size.
Wilkinson (1963) has analyzed this process, and traced the failure
to a too large perturbation of precisely this kind.

The second consequence of (3) is contained in the
phenomenon of "ill-conditioning", which might be called *"inherent
instability"* to emphasize the fact that this is a function of the
problem and not of the method used to solve it. The term ill-conditioned here means that small changes in the coefficients cause
large changes in the zeros, and the effect on the mathematical
problem is that the induced perturbation $\delta_T a_k$, even with stable
methods, and perhaps even the rounding $\delta_M a_k$, are large enough to
prevent the zeros of the perturbed problem (3) from being sufficiently close to the true zeros of the exact mathematical problem.
The phrase "single-precision inadequate" (for example in Professor
Ostrowski's flow diagram) presumably has this interpretation, and
it is then absolutely necessary to work with double or even multi-length arithmetic to ensure that $\delta_T a_k$ and $\delta_M a_k$ are so small that the
computed zeros have the required accuracy.

The mathematical problem has perhaps more interest for
the "mathematical" numerical analyst, but the "practical" numerical
analyst (for example the Director of a University Computing Laboratory!) is probably concerned at least as much with the corresponding
physical problem. Here the coefficients are not known exactly in
advance, but may be subject to errors, say of measurement, of known
upper bound. Equations (1) and (3) are then replaced by the respective equations

(4) $$\sum_{k=0}^{n} (a_k + \delta_P a_k) z^k = 0, \quad \sum_{k=0}^{n} (a_k + \delta_P a_k + \delta_T a_k) z^k = 0,$$

where the subscript P means "physical". Here $\delta_P a_k$ may be much larger than $\delta_T a_k$. For example, in floating-point nested multiplication, we have the bound

(5) $$|\delta_T a_k| \leq (2n+1) 2^{-t} |a_k|,$$

and for a polynomial of degree as large as 100 this would affect only the last (least significant) 8 binary digits of a_k. For a physical problem in which the coefficients are known only to six decimal figures, say, and with a machine which stores the equivalent of ten decimal digits, the four "guarding" digits are quite sufficient to absorb the induced error $\delta_T a_k$.

Now in the mathematical problem (3), which we actually solve, we can think of the data as lying within given intervals, or regions in the complex case, and the zeros lie in corresponding intervals or regions. By using storage and arithmetic of sufficiently high precision we can reduce the size of the data regions as much as we please, and hence also contract each solution region virtually to a point in the complex plane. The physical problem is quite different. We can reduce the $\delta_T a_k$ terms effectively to zero by sufficiently accurate operations, but the physical uncertainties $\delta_P a_k$ are always present. The data regions are always of finite size, and so are the corresponding solution regions. The use of highly-accurate arithmetic can here achieve no more than a sharpening, a focus, of the boundaries of the solution regions.

In the physical problem, therefore, our input should give the data regions and the output should produce the solution regions, from which, for example, we can determine *not* a requested number of accurate figures, which is now virtually meaningless, but the number of *meaningful figures*, the number which is common to all parts of the solution region. An attempt in this direction has been started by R.E. Moore, in his Interval Analysis (1966). He shows that his interval arithmetic fails to satisfy the distributive law, with the consequence that in the evaluation of most functions of interval variables the resulting interval overestimates the true solution interval. The zeros of a polynomial are complicated implicit functions of the coefficients, and the determination of accurate boundaries for the solution regions may require considerable analysis and experiment.

The purpose of this note is to advertise this problem, rather than to attempt its solution, and to stress its importance in practical work. (Since this was written Professor Nickel has indicated that he has made significant advances in this connexion, and has written a successful program of this type). One might further extend the research, to find the probability distribution within the solution region, given the (perhaps different) probability distributions in the data regions. The solution of this problem, and for that matter of allied problems in other areas of practical numerical mathematics, would materially improve the efficiency of our computing services and the value of computer-dependent research.

REFERENCES

R.E. Moore: Interval Analysis, Prentice-Hall, 1966.

J.H. Wilkinson: Rounding Errors in Algebraic Processes, London, H.M.S.O., 1963.

Prof. L. Fox
Oxford University Computing Laboratory
Oxford, England

R. L. Goodstein

A Constructive Form of the Second Gauss Proof of the Fundamental Theorem of Algebra

The constructive form of the second Gauss proof which I shall describe is constructive in the strictest sense of the term, for polynomials with algebraic coefficients. I shall first outline the proof and then fill in the details which show that the operations involved are effective for algebraic numbers.

I start by defining an operation V_k which maps the class of polynomials with algebraic coefficients into itself. If $\alpha_1, \alpha_2, \ldots, \alpha_p$ are the zeros of the polynomial

$$\phi(x) \equiv x^p + a_1 x^{p-1} + a_2 x^{p-2} + \ldots + a_p,$$

the product

$$\prod_{1 \le i < j \le p} \{x - (\alpha_i \alpha_j + k\alpha_i + k\alpha_j)\}$$

is symmetrical in the parameters $\alpha_1, \alpha_2, \ldots, \alpha_p$ and may therefore be expressed as a polynomial $V_k \phi$, of degree $p(p-1)/2$, with coefficients which are polynomials in $a_1, a_2, \ldots, a_p$. The polynomial $V_k \phi$ may be defined directly in terms of the coefficients of ϕ, without any reference to the zeros of $\phi(x)$, as follows.

Let $T_k\phi = \phi(x-k)$, so that the zeros of $T_k\phi$ are $\alpha_i + k$, $1 \leq i \leq p$, and let $E\phi$ be the eliminant of t from the equations $\phi(t) = 0$ and $\phi(x/t) = 0$; $E\phi$ may be expressed as a determinant of order $2p$ in the coefficients of ϕ by Sylvester's method, and is a polynomial in x of degree p^2, with zeros $\alpha_i\alpha_j$, $1 \leq i \leq p$, $1 \leq j \leq p$. Further let $R\phi$ be the polynomial $\phi(\sqrt{x})\,\phi(-\sqrt{x})$ with zeros α_i^2, $1 \leq i \leq p$, so that $E\phi/R\phi$ is the square of a polynomial with zeros $\alpha_i\alpha_j$, $1 \leq i < j \leq p$; for a polynomial f which is a perfect square, $\sqrt{f}$ may be determined by a simple algorithm: for instance if we denote the highest common factor of polynomials ψ_1, ψ_2 by $\{\psi_1, \psi_2\}$ and define $f_{n+1} = \{f_n, f_n'\}^2/f_n$, $f_0 = f$, and $\gamma_{n+1} = f_n/\{f_n, f_n'\}$ then the degree of f_n is strictly decreasing and for some q, f_q is a constant which we may take to be unity and, as is readily seen, $f = (\gamma_1\gamma_2\cdots\gamma_q)^2$ so that $\sqrt{f} = \gamma_1\gamma_2\cdots\gamma_q$. Finally we define $V_k\phi = T_{-k^2}\sqrt{\{ET_k\phi/RT_k\phi\}}$.

For the mapping V_k we have the

<u>Lemma</u>

If we can determine a zero of each of the polynomials $V_k\phi$, $k = 0,1,\ldots,P = p(p-1)/2$, then we can determine a zero of ϕ. For if v_k is a zero of $V_k\phi$ for $k = 0,1,2,\ldots,P$ then two of $v_0, v_1, \ldots, v_p$ must be associated with the same pair α_r, α_s (since there are only P such pairs); enumerate all the pairs (r,s) with $0 \leq r < s \leq P$ in some order and for each pair (r,s) solve the equations

$$\alpha\beta + r(\alpha + \beta) = v_r$$

$$\alpha\beta + s(\alpha + \beta) = v_s$$

to determine $\alpha = \alpha_{rs}$, $\beta=\beta_{rs}$ and test α_{rs} for each pair r,s to find if it is a zero of $\phi(x)$ or not. For some pair r,s necessarily v_r, v_s correspond to the same pair $\alpha_\lambda, \alpha_\mu$ and so α_{rs} is a zero α_λ of $\phi(x)$.

We proceed now to build up a tree of polynomials on a given polynomial $p(x)$ of degree n, with real algebraic coefficients. Let ν be the index of the greatest power of 2 which divides n, let $N_0 = n$, $N_{k+1} = (1/2)N_k(N_k-1)$ so that N_ν is an odd integer, and let $M_0 = 1$, $M_{k+1} = M_k(N_{k+1}+1)$. At the first level above $p(x)$ we place the M_1 polynomials $V_k p$, $k = 0,1,2,\ldots,N_1$, each of degree N_1. If

$$P_1^r, P_2^r, \ldots, P_{M_r}^r$$

are the M_r polynomials of degree N_r at the rth level then $V_i(P_j^r)$, $0 \leq i \leq N_r$; $1 \leq j \leq M_r$ are the M_{r+1} polynomials of degree N_{r+1} at the $(r+1)$th level in the tree. Since the polynomials at the νth level are all of degree N_ν which is odd, we can determine a zero of each polynomial at this level, and hence by the lemma we can determine in turn a zero of each polynomial of each lower level, and so finally we determine a zero of $p(x)$.

It remains to show that the operations in the foregoing proof can be carried out effectively. We begin by showing that if (r_n) is a primitive recursive sequence of rational numbers which is primitively recursively convergent and if $f(x)$ is a polynomial with integral coefficients such that $f(r_n)$ is primitively recursively convergent to zero, so that (r_n) is a zero of $f(x)$, then we can effectively decide whether (r_n) is a zero of a polynomial $g(x)$ with integral coefficients. Let

$$f(x) = a_p x^p + a_{p-1} x^{p-1} + \ldots + a_0$$

and choose an integer $R \geq 1$ so that $R \geq 2(|a_0| + |a_1| + \ldots + |a_{p-1}|)/|a_p|$ then for $|x| \geq R$, $|f(x)| \geq (1/2)|a_p|R^p$; but $f(r_n)$ is primitively recursively convergent to zero and so we can determine an integer ρ so that $|f(r_n)| < (1/2)|a_p|R^p$ for $n \geq \rho$ and therefore $|r_n| < R$ for $n \geq \rho$.

Since f and g have integral coefficients we can effectively decide whether f and g have a common zero or not; if they have no common zero then (r_n) is not a zero of g. If however they have a common zero it remains to be decided whether (r_n) is a common zero or not. Let $h(x)$ be the highest common factor of $f(x)$ and $g(x)$, then we require to decide whether (r_n) is a zero of $h(x)$ or not. Let $f = f_1 h$ and let h_1 be the highest common factor of f_1 and h and write $f_1 = f_2 h_1$, $h = k h_1$; then let h_2 be the highest common factor of f_2 and h_1 and write $f_2 = f_3 h_2$, $h_1 = k_1 h_2$, and so on. In this way, since the degrees of $f_1, f_2, f_3, \ldots$ are strictly decreasing, we arrive at relatively prime F, G where

$$F = f_{m+1}, \quad G = k\, k_1^2\, k_2^3 \ldots k_{m-1}^m\, h_m^{m+1}$$

and
$$h = k\, k_1\, k_2 \ldots k_{m-1}\, h_m$$

and
$$f = FG$$

Since F, G have no common factor, we can find polynomials A, B (with rational coefficients) so that

$$AF + BG = 1 \, .$$

Let a, b, c be upper bounds of $|A|, |B|, |F|$ in the interval $|x| \leq R$. Since $g(r_n)$, $F(r_n)$ and $A(r_n)$ are primitively recursively convergent we can determine $\gamma (\geq \rho)$ so that, for $n \geq \gamma$

$$|G(r_n) - G(r_\gamma)| < 1/5b$$

$$|F(r_n) - F(r_\gamma)| < 1/5a$$

$$|A(r_n) - A(r_\gamma)| < 1/5c.$$

Evaluate the rational number $G(r_\gamma)$; if $|G(r_\gamma)| \geq 2/5b$ then $|G(r_n)| \geq 1/5b$ for $n \geq \gamma$ and (r_n) is not a zero of G; if $|G(r_\gamma)| < 2/5b$ then $|B(r_\gamma) G(r_\gamma)| < \frac{2}{5}$ and so $|A(r_\gamma) F(r_\gamma)| \geq \frac{3}{5}$ whence

$$|A(r_n) F(r_n)| = |A(r_\gamma) F(r_\gamma) + A(r_n)\{F(r_n) - F(r_\gamma)\} + F(r_\gamma)\{A(r_n) - A(r_\gamma)\}|$$

$$\geq \frac{3}{5} - \frac{1}{5} - \frac{1}{5} = \frac{1}{5}.$$

But $F(r_n) G(r_n) = f(r_n)$ is primitively recursively convergent to zero and so given ε we can determine an integer $\delta (\geq \gamma)$ such that $|F(r_n) G(r_n)| < \varepsilon/5a$ for $n \geq \delta$ and therefore $|A(r_n) F(r_n) G(r_n)| < \varepsilon/5$, whence $|G(r_n)| < \varepsilon$ for $n \geq \delta$, which proves that (r_n) is a zero of G; since $G = k \, k_1^2 \, k_2^3 \ldots k_{m-1}^{m} \, h_m^{m+1}$, and $h = k \, k_1 \, k_2 \ldots k_{m-1} \, h_m$ it readily follows that (r_n) is a zero of h and so of g. Correspondingly any zero of h is a zero of G so that when (r_n) is not a zero of G it is not a zero of h.

This completes the proof that we can effectively decide whether a zero of f is a zero of g or not.

Since $|r_n| < R$ for $n \geq \rho$, therefore all r_n lie in some interval i say; let U be an upper bound of $|k_1|, |k_2|, \ldots, |k_{m-1}|, |h_m|$ in i, and $N = m(m+1)/2$ then in i

$$|G(x)| = |h \, k_1 \, k_2^2 \ldots k_{m-1}^{m-1} \, h_m^m| \leq |h| U^N$$

and so if $|G(x)| \geq \alpha$ then $|h| \geq \alpha/U^N$. Moreover, as h is the highest common factor of f, g we can determine polynomials λ, μ such that $\lambda f + \mu g = h$; but we can determine σ so that

$$|\lambda(r_n) f(r_n)| < (1/2) \alpha/U^N \quad \text{for} \quad n \geq \sigma$$

and so $\quad |\mu(r_n) g(r_n)| \geq (1/2) \alpha/U^N$

and finally, taking M as an upper bound for μ in i, we conclude that if $|G(x)| \geq \alpha$ then $|g(r_n)| \geq (1/2)\alpha/MU^N = \eta$, say.

Next we observe that we can effectively decide whether an algebraic number is zero or not. For if (r_n) is a primitively recursively convergent sequence of rationals such that $f(r_n)$ is primitively recursively convergent to zero, where $f(x)$ is the polynomial

$$a_p x^p + \ldots + a_1 x + a_0$$

with integral coefficients, and if $a_0 \neq 0$ then we can determine X so that for $|x| < X$ we have $|f(x)| \geq (1/2)|a_0|$; but we can determine an integer α so that $|f(r_n)| < (1/2)|a_0|$ for $n \geq \alpha$ and therefore $|r_n| \geq X$ for $n \geq \alpha$ and (r_n) does not converge to zero. If $a_0 = a_1 = \ldots = a_t = 0$ and $a_{t+1} \neq 0$, then we determine whether (r_n)

is a zero of $g(x) = a_p x^{p-t-1} + a_{p-1} x^{p-t-2} + \ldots + a_{t+1}$ or not; if it is, then (r_n) does not converge to zero, and if it is not then, as we have shown, we can determine η and for any $\varepsilon > 0$ we can determine δ so that $|g(r_n)| \geq \eta$ and $|r_n^{t+1}(g(r_n))| < \varepsilon^{t+1}\eta$ for $n \geq \delta$, and therefore $|r_n| < \varepsilon$ so that (r_n) converges to zero. When (r_n) does not converge to zero we have seen that we can determine X and α so that $|r_n| \geq X$ for $n \geq \alpha$; by recursive convergence we can determine $\beta \geq \alpha$ so that $|r_n - r_\beta| \leq (1/2)X$ and therefore, if $r_\beta \geq X$ then $r_n \geq (1/2)X$ for $n \geq \beta$, and if $r_\beta \leq -X$ then $r_n \leq -(1/2)X$ for $n \geq \beta$, so that we can effectively determine whether (r_n) is positive or negative.

It follows that if $\phi(x)$ is a polynomial with algebraic real coefficients and if ξ is an algebraic real number, then $\phi(\xi)$ is an algebraic real number, and so we can effectively determine whether $\phi(\xi)$ is zero or not, and if $\phi(\xi) \neq 0$ we can effectively determine whether $\phi(\xi)$ is positive or negative.

Next we require to show that if $\phi(x)$ is a polynomial of odd degree with algebraic real coefficients, then we can determine a primitive recursively convergent sequence of rationals r_n such that $\phi(r_n)$ converges to zero. Let

$$\phi(x) = x^p + \mu_1 x^{p-1} + \mu_2 x^{p-2} + \ldots + \mu_p$$

and let M be an integral upper bound of $1 + |\mu_1| + |\mu_2| + \ldots + |\mu_p|$, so that $\phi(M)$ is positive and $\phi(-M)$ is negative. By repeated bisection we determine a sequence of rational intervals (a_k, b_k) such that $a_0 = -M$, $b_0 = M$, $\phi(a_k)$ is negative and $\phi(b_k)$ is positive and $b_{k+1} - a_{k+1} = (1/2)(b_k - a_k)$; given any $\varepsilon > 0$ the continuity of ϕ determines a δ so that, for $k \geq \delta$,

$$\phi(b_k) - \phi(a_k) < \varepsilon$$

and therefore $\phi(b_k) < \varepsilon$, proving that the primitive recursively convergent sequence of rationals (b_n) is a root of $\phi(x)$.

Finally we consider a polynomial $\Phi(x)$, of degree n, with algebraic complex coefficients, so that $\Phi(x) = F(x) + iG(x)$, where F,G are polynomials of degree n (at most) with real algebraic coefficients. If F and G have a non-constant common factor $H(x)$, then a zero of $H(x)$ is a root of $\Phi(x)$; if F and G are relatively prime we consider $\Psi(x) = \{F(x) + iG(x)\}\{F(x) - iG(x)\}$, a polynomial of degree $2n$. $F(x) + iG(x)$ and $F(x) - iG(x)$ have no common factor. We determine $n+1$ zeros of $\Psi(x)$ as follows. $\Psi(x)$ has no real zero, since a real zero of Ψ would be a zero of both F and G. We determine first a complex zero of Ψ, and then divide Ψ by the corresponding quadratic factor with real coefficients, and determine a zero of the quotient, and so on. At least one of the $n+1$ zeros of Ψ is a zero of $F(x) + iG(x)$ and so a zero of $\Phi(x)$.

Prof. R.L. Goodstein
Department of Mathematics
The University
Leicester, England

Peter Henrici and Irene Gargantini

Uniformly Convergent Algorithms for the Simultaneous Approximation of all Zeros of a Polynomial

Abstract

A practical algorithm, based on H. Weyl's proof of the fundamental theorem of algebra, is presented for the simultaneous determination of all zeros of a given polynomial with complex coefficients. The algorithm is uniformly convergent in the space of all polynomials of a given degree. No starting approximations are required. Four versions of the algorithm are indicated, one of them possessing uniform linear convergence. An implementation of the algorithm in an experimental computer program written for the IBM 360 model 40 is described, and some typical results are discussed.

PART A: THEORY

1. Introduction

In many well-known algorithms for calculating the zeros of a polynomial, linear or quadratic factors are determined one at a time. If an approximate factor has been obtained, the polynomial is divided by it, and the same algorithm is then applied to the reduced polynomial. For instance, the algorithms reported in the papers [8], [1], [10], [3] are of this type. In spite of their widespread use, however, algorithms of this type have some theoretical and practical shortcomings. The theoretical flaw consists in the fact that these algorithms, even if they produce each single factor of the polynomial as limit of a primitive recursive sequence, do not produce the set of *all* factors as such a limit, at least as long as the cumulative effect of dividing out inaccurate factors is not analyzed very carefully. From the practical point of view this leads to the disadvantage that even in a situation where only moderately accurate zeros are required, all factors except the last have to be determined to full working accuracy in order not to spoil the late factors completely by dividing out first factors of moderate accuracy.

In the present paper we shall consider the problem of simultaneously calculating all zeros of a polynomial with an error guaranteed not to exceed a preassigned value. No special hypotheses about the polynomial are made, and we do not suppose that "sufficiently close" approximations to the zeros are known. In these respects our algorithm differs from other methods for the simultaneous determination of all zeros such as the quotient-difference algorithm of Rutishauser [12], where it is assumed that the zeros have distinct moduli, or from the algorithm of Weierstrass [14] as recently revived by Kerner [6], where the zeros are supposed to be simple, and where close first approximations to the zeros are required.

The precise problem to be solved is as follows. Let there be given a polynomial P of degree $N > 0$ with complex coefficients, a number $D > 0$ with the property that all zeros of P lie in a disk of radius D about the origin, and a number η such that $0 < \eta < 1$. We wish to construct a set containing all zeros of P which consists of at most N components, each of diameter $\leq \eta D$, and to specify the number of zeros in each component. Such a set will be called an η-*inclusion set*. An η-inclusion set will be said to determine the zeros with uncertainty η.

An algorithm for constructing η-inclusion sets will be called an *inclusion algorithm*. It will be called *convergent* if it permits the construction of an η-inclusion set for every η > 0. It will be called *uniformly convergent* if the number of arithmetic operations required to construct an η-inclusion set for a given polynomial P is bounded by a quantity ν depending on η and N, but not on P. The existence of such uniformly convergent inclusion algorithms is implied by the results of Specker [13]. Here our concern is with inclusion algorithms that are computationally feasible, and with the dependence of their functions ν on η and N.

A general form of an inclusion algorithm will be described in Section 2. Following an idea of Weyl [15], this algorithm is based on a test T, called exclusion test, which can be applied to any polynomial P and to any square in the complex plane. If the test is passed, this means that the square does not contain any zeros of P. The algorithm consists in systematically subdividing a square known to contain all the zeros and applying T to subsquares of squares which have not passed the test. Some conditions on T will be stated which are sufficient to guarantee the uniform convergence of the algorithm.

Several possible implementations of exclusion tests will be described in Section 3. The first test, T_1, merely requires the evaluation of the polynomial, and no derivatives. It is shown to define a uniformly convergent inclusion algorithm requiring no more than

$$16 \, N \, \pi \left(\frac{2^{5/2} N}{\eta} \right)^{2N-2}$$

evaluations of P to achieve uncertainty η. If the zeros of P are distinct, the number of evaluations is $O(log_2 \eta^{-1})$, that is, the convergence is ultimately linear. Two more exclusion tests, T_2 and T_3, are described next. These require the evaluation of derivatives, but the convergence will be somewhat improved. Finally we shall describe an exclusion test based on the Schur-Cohn algorithm for determining the number of zeros of a polynomial within a given circle. This test, although arithmetically more complicated than the others, will be shown to produce always linear convergence. In fact, the number of applications of the test required to achieve uncertainty η is shown not to exceed

$$8 \pi N \, log_2 \, \frac{4 N}{\eta} \, .$$

In order not to obscure the presentation with computational detail, questions of round-off error are not considered in the present paper. It will be clear, however, that our tests and algorithms could also be formulated in the presence of round-off. Naturally, the elementary round-off error to be tolerated to achieve a given uncertainty $\eta > 0$ will be a function of η. We conjecture that this function is of the form $c \eta^N$ where c is independent of N, $0 < c \leq 1$. For $N = 2$ this result (with $c = 1$) was recently established by Kahan [4].

2. A General Inclusion Algorithm

Without loss of generality we may assume that D, the diameter of the initial disk known to contain all zeros of P, equals 2, since this can always be achieved by a trivial substitution. We also may assume that the leading coefficient of the given polynomial is 1. We thus denote by $\mathbf{P}_N$ the class of all polynomials $P(z) = z^N + a_{N-1} z^{N-1} + \ldots + a_0$ whose zeros $z_1, z_2, \ldots, z_N$ satisfy $|z_i| < 1$, $i = 1, 2, \ldots, N$. Our problem then is to construct, for any $P \in \mathbf{P}_N$ and any $\eta > 0$, a set S with $n \leq N$ components $C^1, C^2, \ldots, C^n$, each having a diameter $< 2\eta$, such that all zeros of P are contained in S, and to determine the precise number of zeros in each C^i.

2.1 The exclusion test

We suppose that there is at our disposal a test T which is applicable to any $P \in \mathbf{P}_N$ and to any closed square Q of the complex plane, and which has the following property: If Q passes the test, then no zero of P is contained in Q. A square which does not pass the test is called *suspect*. It is not required that a suspect square actually contain a zero. Any test with the above property will be called an exclusion test. In order to be useful an exclusion test must have several additional properties which will be discussed presently.

We now define a nested sequence of inclusion sets S_0, $S_1, \ldots$, as follows. Let Q_0 denote the closed square

$$|\text{Re } z| \leq 1, \quad |\text{Im } z| \leq 1.$$

Since Q_0 contains all zeros of P, Q_0 is suspect. We let $S_0 = Q_0$. We now divide Q_0 into four congruent subsquares Q_1 by joining midpoints of opposite sides by straight lines. (The joining lines belong to either adjacent square.) To each of the resulting subsquares we apply the test T. The set S_1 is the union of the suspect squares Q_1. We now continue in the obvious manner: Having defined a set S_h of suspect squares Q_h of edge 2^{1-h}, we divide each $Q_h \subset S_h$

into four congruent subsquares Q_{h+1} of sidelength 2^{-h} to which we apply the criterion T. The set S_{h+1} is the union of all suspect squares Q_{h+1}.

Let the zeros of P be denoted by $z_1, z_2, \ldots, z_N$. Clearly, every $z_i \in S_h$ for $h = 0, 1, 2, \ldots$. On the other hand, if we define

(2-1) $$\varepsilon_h^{T,P} = \sup_{z \in S_h} \min_{1 \leq i \leq N} |z - z_i| ,$$

then the set S_h is contained in the N disks of radius $\varepsilon_h^{T,P}$ about the points z_i, $i = 1, \ldots, N$. We now define the *convergence function of the test T* by

(2-2) $$\varepsilon_h^T = \sup_{P \in \mathbf{P}_N} \varepsilon_h^{T,P} .$$

The test is called *convergent* if

$$\lim_{h \to \infty} \varepsilon_h^T = 0 .$$

For a convergent test, the algorithm described above results in sets S_h that are known to contain all zeros and that are contained in the union of at most N disks of arbitrarily small radius.

2.2 Pregnant components

In general the set S_h consists of several components (i.e., maximal connected subsets) $C_h^1, \ldots, C_h^m$ where $m > 1$. By the covering property mentioned above, the diameter of each component cannot exceed $2N \varepsilon_h^{T,P}$. Thus if T is convergent the diameter of each component tends to zero. Nevertheless the problem of determining the zeros with arbitrarily small uncertainty $\eta > 0$, as stated in the Introduction, will not be solved unless we also determine the number of zeros in each component.

Still following Weyl, this can be achieved constructively by a discrete version of the argument principle. For this to be applicable, the test T must have the following additional property: If the square Q is not suspect, the set $P(Q)$ lies in the interior of a suitable half-plane bounded by a straight line through the origin. A test with this property will be called *one-sided*.

Let C denote a component of S_h, and let Γ denote its boundary, oriented such that C lies to the left of Γ. If C is multiply connected, Γ consists of several Jordan curves, $\Gamma_0, \Gamma_1, \ldots, \Gamma_I$. By the principle of the argument the number of zeros of P in C is given by

$$(2\text{-}3) \qquad n = \frac{1}{2\pi} \sum_{i=0}^{I} [\arg P(z)]_{\Gamma_i},$$

where the brackets denote the variation of the argument along the closed curve Γ_i. To evaluate the sum, we recall that C is the union of certain suspect squares Q_h. Consequently, each Γ_i consists of a finite number of sides of squares Q_h, i.e. of straight line segments σ_{ik} ($k = 1, 2, \ldots, K_i$) of length 2^{1-h}, joined together at the points $z_{i0}, z_{i1}, \ldots, z_{iK_i} = z_{i0}$. We consequently have

$$(2\text{-}4) \qquad [\arg P(z)]_{\Gamma_i} = \sum_{k=1}^{K_i} [\arg P(z)]_{\sigma_{ik}}.$$

Now each σ_{ik}, being part of the boundary of C, is also part of a side of a certain non-suspect square.[*] If the test T is one-sided,

---

[*] Except for small values of h, since then σ_{ik} may coincide with part of the boundary of Q_0.

it follows that the set $P(\sigma_{ik})$ lies entirely on one side of a suitable straight line through O, and hence that

$$\left|[\arg P(z)]_{\sigma_{ik}}\right| < \pi, \quad k = 1, 2, \ldots, K_i.$$

The sum appearing in (2-4) thus may be evaluated recursively by selecting an arbitrary value of $\arg P(z_{i,0})$ and, having determined $\arg P(z_{i,k-1})$, choosing for $\arg P(z_{i,k})$ the unique value satisfying

$$\left|\arg P(z_{i,k}) - \arg P(z_{i,k-1})\right| < \pi.$$

With this selection of the arguments,

(2-5) $\qquad [\arg P(z)]_{\Gamma_i} = \arg P(z_{i,K_i}) - \arg P(z_{i,0}).$

2.3 Simplification of the suspect sets

It is evident that the bookkeeping necessary to keep track of the boundaries of the components of the sets S_h can become enormously involved, at least potentially. To simplify the algorithm of Section 2.2 we shall replace the set S_h by a set S_h^* whose components have simple boundaries, without endangering the convergence of the method. To describe the process, we shall denote, for any compact set C, by $R(C)$ the circumscribed rectangular set whose boundaries run parallel to the coordinate axes.

To replace S_h by a simpler set, we construct a sequence of sets $S_h^j (j = 0, 1, \ldots)$, whose components we denote by $C_h^{j,m} (m = 1, 2, \ldots, M_j)$, as follows: Let $S_h^0 = S_h$, and for $j = 0, 1, \ldots$

(2-6) $$S_h^{j+1} = \bigcup_{m=1}^{M_j} R(C_h^{j,m}).$$

Each S_h^{j+1} is thus obtained by replacing the components of S_h^j by their circumscribed rectangles. Replacing the components by larger connected sets can only increase their connectedness, hence the number of components cannot increase. It can decrease if some of the circumscribed rectangles overlap. However, since $M_j \geq 1$, there exists a smallest j such that $M_j = M_{j+1}$. The set $S_h^* = S_h^{j+1}$ then is the union of M_j non-overlapping rectangles. The algorithm of Section 2.2 can now be used to determine the number of zeros in each rectangular component. Since the original set S_h can be covered by at most N disks of radius ε_h^T, the diameter of each rectangular component cannot exceed $\delta_h^T = 2\sqrt{2} N \varepsilon_h^T$. Thus if the test T is convergent and one-sided, the algorithms described in Section 2.1 - 2.3 solve the problem stated in the Introduction.

The experimental program described in Part B lists the centers of the rectangular components R of S_h^* as approximations to the zeros of P, and the semidiagonals of R as an upper bound for the error of the approximation. Each center is listed as many times as there are zeros in R.

2.4 A m o u n t   o f   w o r k

If an upper bound for the convergence function ε_h^T is known, it is easy to estimate the number of applications of T that are necessary to construct S_h. Since each S_k can be covered by N disks of radius ε_k^T, its total area is at most $N\pi(\varepsilon_k^T)^2$, and it thus cannot contain more than

$$2^{2k} N\pi (\varepsilon_k^T)^2$$

squares Q_{k+1} of edge 2^{-k} to which the test must be applied in order to construct S_{k+1}. If ν_h^T denotes the total number of applications of T necessary to construct S_h, we thus have

(2-7) $$\nu_h^T \leq N \pi \sum_{k=0}^{h-1} 4^k (\varepsilon_k^T)^2.$$

3. Some Special Exclusion Tests

In this section we shall present several examples of exclusion tests. These tests are all convergent and, with one exception, one-sided. In some instances we shall compare the effectiveness of different tests. A test T is said to be at least as effective as a test T' if, for any polynomial P and any square Q, Q is suspect by T' whenever it is suspect by T. If T is at least as effective as T', then clearly

(3-1) $$\varepsilon_h^T \leq \varepsilon_h^{T'}, \quad h = 0,1,2,\ldots$$

3.1 The test T_1

This test depends on a constant K, $T_1 = T_1(K)$. If $P \in \mathbf{P}_N$ and if Q is a square with center a and semidiagonal r, we call Q suspect according to $T_1(K)$ if

(3-2) $$|P(a)| \leq K r.$$

We define

(3-3) $$K_N = N(1 + \sqrt{2})^{N-1}$$

and state

Theorem 3.1

$T_1(K)$ *is a one-sided, uniformly convergent exclusion test for all* $K \geq K_N$. *If* $N \geq 2$ *and* $K = K_N$, *then*

(3-4) $$\varepsilon_h^{T_1} \leq 4 \cdot 2^{-\frac{h}{N}}, \quad h = 0,1,2,\ldots .$$

The significance of the constant K_N is explained by the following

Lemma 3.1

(3-5) $$\sup_{P \in \mathbf{P}_N} \sup_{z \in Q_0} |P'(z)| = K_N.$$

P r o o f of the Lemma. Let

$$P'(z) = N \prod_{i=1}^{N-1} (z - z_i') .$$

By a well-known result of Gauss (see Ref. [11], chapter 3, problem 31) the zeros of P' lie in the convex hull of the zeros of P, hence $|z_i'| < 1$, $i = 1,2,\ldots,N-1$. It is geometrically evident that

$$\sup_{\substack{z \in Q_0 \\ |z_i'| < 1}} |z - z_i'| = 1 + \sqrt{2} .$$

Hence the supremum on the left of (3-5) cannot exceed K_N. Consideration of the polynomials

$$P(z) = (z - te^{i\frac{\pi}{4}})^N,$$

when t is arbitrarily close to 1, shows that equality holds.

P r o o f of the Theorem. Let $K \geq K_N$. If z is any point of Q, then by Lemma 3.1, since $|z-a| \leq r$,

$$\left|P(z) - P(a)\right| = \left|\int_a^z P'(t)\, dt\right| \leq K_N r \leq Kr.$$

Hence, if Q is not suspect, then $P(z)$ is contained in a circle around $P(a)$ with radius $< |P(a)|$. Hence $T_1(K)$ is a one-sided exclusion test.

To show that $T_1(K)$ is convergent, we require the following simple observation. *Let P be any polynomial of degree N and leading coefficient 1, and let $\varepsilon \geq 0$. If the point a is such that $|P(a)| \leq \varepsilon$, then*

$$|a - z_i| \leq \sqrt[N]{\varepsilon}$$

for at least one zero z_i of P. Speaking intuitively, this means that a polynomial with leading coefficient 1 cannot be small far away from any zero. To prove the observation, suppose that

$$|a - z_i| > \sqrt[N]{\varepsilon}$$

for all zeros z_i of P. Then

$$|P(a)| \prod_{i=1}^{N} |a-z_i| > (\sqrt[N]{\varepsilon})^N = \varepsilon,$$

contradicting the hypothesis.

Applying the observation with $\varepsilon = Kr$, we find that for $h = 0, 1, 2, \ldots$ the center of any suspect square Q_h (where $r = 2^{1/2-h}$) is at a distance not exceeding

$$(\sqrt{2}\, K)^{1/N}\, 2^{-\frac{h}{N}}$$

from a suitable zero of P. It thus follows that

$$\varepsilon_h^{T_1(K)} \leq (\sqrt{2}\, K)^{1/N} + 2^{1/2-h}.$$

This clearly converges to zero for $h \to \infty$, proving $T_1(K)$ to be convergent. If $K = K_N$, an elementary computation shows that

(3-6) $$(\sqrt{2}\, K_N)^{1/N} = (1+\sqrt{2})\left(\frac{\sqrt{2}\, N}{1+\sqrt{2}}\right)^{1/N} < 3.$$

Since $2^{1/2-h} \leq 2^{-\frac{h}{N}}$ for $h \geq 1$ and $N \geq 2$, the inequality (3-4) now clearly follows for $h \geq 1$, while it is trivial for $h = 0$. This completes the proof of Theorem 3.1.

An application of the test T_1 essentially involves one evaluation of P and thus is clearly independent of h. If $K = K_N$ we find from (2-7), using (3-4), that

$$\nu_h^{T_1} \leq 16\, N\pi \sum_{k=0}^{h-1} 2^{(2-\frac{2}{N})k} \leq 16\, N\pi 2^{(2-\frac{2}{N})h}.$$

To achieve uncertainty η in the location of the zeros, it is necessary that the quantity $\delta_h^{T_1}$ defined in Section 2.3 satisfy $\delta_h^{T_1} \leq 2\eta$. This will be the case if $\sqrt{2N}\, \varepsilon_h^T \leq \eta$, or, using (3.4), if

$$h \geq N(\log_2 \frac{N}{\eta} + \frac{5}{2}) ,$$

which requires at most

(3-7) $$16\, N\, \pi\, (\frac{2^{5/2} N}{\eta})^{2N-2}$$

applications of the test. Since this bound depends only on η and N, we have shown that the test T_1 does indeed produce a uniformly convergent algorithm.

3.2 Separated zeros

Although Theorem 1 proves uniform convergence, the number of operations required to achieve a given uncertainty η was shown to be bounded only by $O(\eta^{2-2N})$ which is intolerably large for numerical purposes. We shall now show that the slow speed is due to possible clusters of zeros, and that much better estimates for the speed of convergence can be given if something is known about the separation of the zeros.

Theorem 3.2

Let $N \geq 2$, let $P(z) = z^N + \ldots$ have the distinct zeros $z_1, z_2, \ldots, z_n$ $(n \leq N)$, let μ be the maximum multiplicity of any z_i, and let

$$\Delta = \min_{z_i \neq z_j} |z_i - z_j| .$$

If h is such that

(3-8) $$[1 + (\sqrt{2}\ K)^{1/N}]\ 2^{-\frac{h}{N}} < \frac{1}{2} \Delta ,$$

then

(3-9) $$\varepsilon_k^{T_1(K),P} \leq A \cdot 2^{-\frac{h}{\mu}} ,$$

where A is a constant depending on K, N, μ and Δ.

P r o o f

Let a be the center of a suspect Q_h. By the proof of Theorem 1, $|a-z_i| \leq [1+(\sqrt{2}\ K)^{1/N}]\ 2^{-h/N}$ for at least one zero z_i. However, in view of (3-8) this can only hold for precisely one index i, say for $i = j$, and $|a-z_k| > 1/2\Delta$ for $k \neq j$. If $|P(a)| \leq \varepsilon$ it follows from

$$\varepsilon \geq |P(a)| \geq |a-z_j|^\mu\ (\tfrac{1}{2}\Delta)^{N-\mu}$$

that

$$|a-z_j| \leq \varepsilon^{\frac{1}{\mu}}\ (\tfrac{1}{2}\Delta)^{1-\frac{N}{\mu}} .$$

Hence if $\varepsilon = 2^{1/2-h}K$, then

$$|a-z_j| \leq (2^{\frac{1}{2}} K)^{\frac{1}{\mu}}\ (\tfrac{1}{2}\Delta)^{1-\frac{N}{\mu}}\ 2^{-\frac{h}{\mu}} .$$

Adding to this the length of the semidiagonal of Q_h and bounding it by $2^{1/2-h/\mu}$, we find

$$\varepsilon_h^{T_1(K),P} \leq \left[2^{\frac{1}{2}} + (2^2 K)^{\frac{1}{\mu}} \left(\frac{2}{\Delta}\right)^{\frac{N}{\mu}-1} \right] 2^{-\frac{h}{\mu}},$$

which is of the form (3-9). For $K = K_N$ this can, by virtue of (3-6), be stated in the form

(3-10) $\quad \varepsilon_h^{T_1(K_N),P} \leq 4.45 \left(\frac{6}{\Delta}\right)^{\frac{N}{\mu}-1} 2^{-\frac{h}{\mu}}.$

In the remainder of this section we assume $K = K_N$. Let h_0 be the smallest h satisfying (3-8). By Section 3.1, no more than

(3-11) $\qquad\qquad 16 \, N \, \pi \, \left(\frac{2^{7/2} N}{\Delta}\right)^{2N-2}$

applications of T_1 are needed to construct S_{h_0}. From then onward, using (2-7), only

(3-12) $\qquad\qquad N \pi \sum_{k=h_0}^{h-1} 4^k \left(\varepsilon_k^{T_1,P}\right)^2$

$$= 20 \, N \, \pi \, \left(\frac{6}{\Delta}\right)^{\frac{2N}{\mu}-2} \sum_{k=h_0}^{h-1} 2^{(2-\frac{2}{\mu})k}$$

further applications are needed to construct S_h. If $\mu > 1$, the sum is bounded by

$$2^{(2-\frac{2}{\mu})h}.$$

To achieve uncertainty η, it suffices that $\sqrt{2}\, N\varepsilon^{T_1,P} \leq \eta$. By (3-9) this will be the case if

$$2^h \geq (\frac{7N}{\eta})^\mu \, (\frac{6}{\Delta})^{N-\mu}$$

and a short computation shows that this can be achieved by

$$20\, N\, \pi\, (\frac{7N}{\eta})^{2\mu-2} (\frac{6}{\Delta})^{2N-2\mu} = O(\eta^{2-2\mu})$$

applications of T_1 in addition to (3-11). This represents an improvement over (3-7) if $\mu < N$. In the important case $\mu = 1$ (all zeros simple) the sum in (3-12) equals $h - h_0$, and a similar computation shows that now only

$$20\, N\, \pi\, (\frac{6}{\Delta})^{2N-2} \left[\log_2 \frac{1}{\eta} + \log_2 7\, N(\frac{6}{\Delta})^{N-1} \right] = O(\log \frac{1}{\eta})$$

applications are needed in addition to (3-11). Thus in the case of N distinct zeros the convergence of the algorithm is *linear*.

3.3 The test T_2

Let the square Q have the center a and the semidiagonal r. In order to apply T_2 to the polynomial $P(z) = z^N + \ldots$ we determine the coefficients b_k in the expansion

$$P(z) = \sum_{k=0}^{N} b_k (z - a)^k .$$

Q is called suspect according to T_2 if

(3-13)
$$|b_0| \leq \sum_{k=1}^{N} |b_k|\, r^k .$$

Theorem 3.3

T_2 is a one-sided, uniformly convergent exclusion test.

P r o o f

If Q is not suspect, $z \varepsilon Q$, then

$$\left| P(z) - b_0 \right| = \left| \sum_{k=1}^{N} b_k (z-a)^k \right| \leq \sum_{k=1}^{N} |b_k| r^k < |b_0|.$$

This shows that the set $P(Q)$ lies in a circle around b_0 with radius $< |b_0|$ which already shows that T_2 is a one-sided exclusion test. In order to show that T_2 is convergent, we note that the expression on the right of (3-13) is of the form

$$r \left[|b_1| + |b_2| r + \ldots + |b_{N-1}| r^{N-2} + r^{N-1} \right].$$

In Lemma 1 it was shown that for the class of polynomials under consideration,

$$\sup |b_1| = N(1+\sqrt{2})^{N-1} = K_N.$$

In a completely analogous manner we can show that

(3-14) $$\sup |b_k| = \binom{N}{k} (1 + \sqrt{2})^{N-k}, \quad k = 1, 2, \ldots, N-1.$$

Since $r_h \leq \sqrt{2}$, the expression in brackets is thus bounded by $(1+2\sqrt{2})^N$. It follows that the test T_2 is at least as effective as the test $T_1(K)$ where

$$K = (1 + 2\sqrt{2})^N,$$

hence convergent. Although the convergence estimates do not show it, the test T_2 is asymptotically likely to be much more effective than $T_1(K_N)$, since the constant $|b_1|$, which is the only one that matters for small values of r, is likely to be much smaller than K_N in most cases.

Each application of T_2 requires the computation of the complete Horner scheme of P for the point $z = a$.

3.4 The test T_3

Let r, a, b_k $(k=0, 1, \ldots, N)$ be defined as above. The square Q is suspect according to T_3 if

$$(3\text{-}15) \qquad |b_0| \leq \sum_{k=1}^{N} k |b_k| r^k .$$

Clearly, this test is less effective than T_2. This already implies that T_3 is a one-sided exclusion test. Using the bounds (3-14) for the b_k, it is seen that T_3 is more effective than the test $T_1(K)$ where

$$K = N(1 + 2\sqrt{2})^{N-1} ,$$

hence convergent. We thus have proved:

Theorem 3.4

T_3 *is a one-sided, uniformly convergent exclusion test.*

The advantage of T_3 over T_2 lies in the possibility that it can be modified as follows. We write the test in the form

$$(3\text{-}16) \qquad |P(a)| \leq r_h K_h(a) ,$$

where

$$K_h(a) = \sum_{k=1}^{N} k |b_k| r_h^{k-1}.$$

Clearly, $K_h(a)$ is an upper bound for $|P'(z)|$ in Q_h. Thus if the square Q_{h+s} with center c is contained in Q_h, the test based on the inequality

(3-17) $$|P(c)| \leq r_{h+s} K_h(a)$$

is again a one-sided exclusion test, for if (3-17) is not satisfied and $z \in Q_{h+s}$, we have

$$|P(z) - P(c)| = \left| \int_c^z P'(t) dt \right| \leq r_{h+s} K_h(a) ,$$

showing that $P(Q_{h+s})$ lies in a circle around $P(c)$ with radius $< |P(c)|$. The modified test remains convergent, for it is still more effective than a suitable $T_1(K)$. However, while T_2 requires the computation of the complete Horner scheme at each step, T_3 requires the computation of the complete scheme only if a new value of $K_h(a)$ is desired. At the other steps only the evaluation of $P(c)$ is needed.

In Fig. 1 we show a typical square Q_h in the z-plane and its image under the transformation $w = P(z)$. Also shown are three circles

$$|w - P(a)| = R^{(i)}, \quad i = 1, 2, 3,$$

where $R^{(i)}$ is the "exclusion radius" determined by the test T_i.

$$R^{(1)} = r\, K_N$$

$$R^{(2)} = r \sum_{k=1}^{N} |b_k| r^{k-1}$$

$$R^{(3)} = r \sum_{k=1}^{N} k|b_k| r^{k-1}.$$

Since all three circles enclose the origin, it turns out that Q_h is suspect according to all three tests, although it evidently does not contain a zero.

3.5 The test T_4

This is a test of a different type which is based on the fact that it is possible by algebraic means to decide whether or not a given polynomial has any zeros in a given circle. Let Q again be a square with center a and semidiagonal r. If

$$P(z) = \sum_{n=0}^{N} a_n z^n \qquad (a_N = 1)$$

is the given polynomial, we set

$$\zeta = \frac{z - a}{r}$$

and

(3-18) $$R_0(\zeta) = P(z) = P(a+r\zeta) = \sum_{n=0}^{N} c_n^{(0)} \zeta^n$$

so that, in the notation of Section 3.3,

$$c_n = b_n r^n, \; n = 0,1,\ldots,N.$$

It will be convenient to employ the following notation. If

$$R(\zeta) = \sum_{n=0}^{M} c_n \zeta^n$$

is any polynomial, we set

$$R^*(\zeta) = \zeta^M \overline{R(\bar{\zeta}^{-1})} = \sum_{n=0}^{M} \bar{c}_{M-n} \zeta^n.$$

Beginning with R_0 as defined by (3-18), the polynomials

$$R_j(\zeta) = \sum_{n=0}^{N-j} c_n^{(j)} \zeta^n$$

are now defined by

(3-19) $$R_{j+1}(\zeta) = \bar{c}_0^{(j)} R_j(\zeta) - c_{N-j}^{(j)} R_j^*(\zeta),$$

$$j = 0,1,2,\ldots$$

or, equivalently

(3-20) $$c_n^{(j+1)} = \bar{c}_0^{(j)} c_n^{(j)} - c_{N-j}^{(j)} \bar{c}_{N-j-n}^{(j)},$$

$$n = 0,1,\ldots, N-j-1; \; j = 0,1,\ldots .$$

It is not asserted that R_j has exact degree $N-j$. However, the following result holds:

Lemma 3.5

The polynomial $R(\zeta)$ does not have any zeros ζ satisfying $|\zeta| \leq 1$ if and only if

(3-21) $\qquad c_0^{(0)} \neq 0, \quad c_0^{(j)} > 0, \quad j = 1, 2, \ldots, n$.

The result is essentially due to Schur and Cohn. It is closely related to (although not identical with) Theorem 42.1 of Ref. [9] (see also Ref. [7]) and can be proved by similar methods.[*]

We call the square Q suspect according to T_4 if the condition (3-21) is not satisfied. It follows from the Lemma that if a square Q_h is suspect, then P has a zero at a distance $\leq r_h = 2^{\sqrt{2}-h}$ from the center of Q_h. We thus have:

Theorem 3.5

T_4 *is a uniformly convergent exclusion test, in fact,*

(3-22) $\qquad \varepsilon_h^{T_4} \leq 2^{\frac{3}{2}-h}, \quad h = 0, 1, 2, \ldots$.

It follows from (3-22) that the number of applications of T_4 required to construct S_h does not exceed

---

[*] The algorithm described by (3-18), (3-19) and (3-20) has been used by Lehmer [7] to determine zeros of polynomials. In Lehmer's version the objective is to decide whether a given open disk contains at least one zero of P. As pointed out in Ref. [7], there are exceptional situations where the algorithm fails in that decision. In our version the objective is to decide whether a given closed disk is free of zeros. In this decision the algorithm (excepting rounding errors) never fails.

$$N \pi \sum_{k=0}^{h-1} 4^k \left(\varepsilon_k^{T_4} \right)^2 = 8 N \pi h .$$

Uncertainty η will be achieved if $\sqrt{2} \, N \, \varepsilon_h^{T_4} < \eta$, i.e. if $2^{-h} \leq \eta/4N$, hence the number of applications of T_4 required to obtain uncertainty η is not larger than

$$8 N \pi \left[2 + \log_2 N + \log_2 \eta^{-1} \right] .$$

Thus the exclusion algorithm based on T_4 is uniformly linearly convergent. In comparing these results to those proved for the other tests, it should be kept in mind that T_4 requires at most $5N^2 + O(N)$ real multiplications compared to $2N^2$ for T_2 or $4N$ for T_1 and the modified form of T_3.

It is not asserted that T_4 is one-sided, and it can be shown by simple examples that in fact it is not. Thus if the set S_h^k has less than N components, and if it is desired to determine the number of zeros in each component the test T_4 must be followed by at least one application of the tests described above, which may result in an enlargement of the set S_h. In practice it has been found useful to use the test T_4 until the zeros are separated, and then to continue with the modification of T_3 described at the end of Section 3.4.

PART B: EXPERIMENTS

4. The Experimental Program

A program has been written for IBM System/360, model 40 in collaboration with W. Münzner[†] with the purpose of testing the numerical performance of the various algorithms considered in PART A. A simplified flow diagram of the program is shown in Fig.2. Some explanations of the flow diagram follow.

BOX 1.

The input consists of $N+1$ short or long precision complex numbers $A_0, A_1, \ldots, A_N$, the coefficients of the polynomial

(4.1) $$P_0(z) = A_N z^N + A_{N-1} z^{N-1} + \ldots + A_0$$

whose zeros $z_1, z_2, \ldots, z_N$ are desired. These zeros are known to lie inside a circle of radius

(4.2) $$R = 2 \max_{1 \leq k \leq N} \left| \frac{A_{N-k}}{A_N} \right|^{1/k}.$$

If $R > 0$, the zeros $z_1, z_2, \ldots, z_N$ of the polynomial

$$P(z) = \frac{P_0(Rz)}{R^N A_N} = z^N + a_{N-1} z^{N-1} + \ldots + a_0$$

where

$$a_k = \frac{A_k}{A_N R^{N-k}}, \quad k=0,1,\ldots,N-1,$$

[†] Presently at IBM Thomas J. Watson Research Center, P.O. Box 218, Yorktown Heights, N.Y. 10598.

thus lie inside the unit circle. The algorithm works with the polynomial P. The user has the option of using any number R in place of (4.2) for computing the normalized polynomial P; in particular if the zeros are already known to lie in $|z| < 1$, one can put $R = 1$.

BOX 3.

The program provides a choice of the four different tests $T_1(K_N)$, T_2, T_3, T_4. T_1 and T_4 can be used only at the beginning, and only for consecutive values of h. At the last step the program always uses T_2, independently of the test used before. If T_1 or T_3 is used and the number of suspect squares exceeds a quarter of the maximum number of squares that can be stored, the program switches to T_2.

BOX 5.

At each step h a list of the coordinates of the centers of all suspect squares and of upper bounds for the absolute value of the first derivative in each suspect square is constructed. Initially the bounds are taken to be K_N, as soon as available, they are replaced by the bounds obtained from the computation of a complete Horner scheme (i.e. by applying T_2 or T_4). The maximum number of suspect squares is 1000 for short and 500 for long precision. The precision used in applying the test is the same as the precision in which the coefficients of P_0 are given. Since the real and imaginary parts of z are negative powers of 2, they are always represented exactly in the machine (for $h \leq 20$ in short precision, and in long precision for $20 < h \leq h_{max}$). To avoid false decisions due to round-off error, the quantities at the right of (3-13) and (3-15) are multiplied by 1.001. For the same reason, when using T_4, the radius $r = 2^{1/2-h}$ of the circle to which the Schur-Cohn criterium is applied is replaced by 1.5×2^{-h}, and, to avoid underflow, the successive polynomials R_j are normalized by dividing by the constant coefficient.

BOX 8

h_{max} is taken to be 24. The minimum sidelength of a square is $2^{-23} \sim 1.2 \times 10^{-7}$.

BOX 9

The construction of the rectangles is performed by the algorithm described in Section 2.3.

BOX 10

See Section 2.2. This computation is always performed in long precision.

BOX 11

If the variation of the argument of P along the boundary of a rectangle is $2\pi p$, the center of the rectangle is accepted as a p-fold approximation to p zeros within the rectangle. The length of the semidiagonal is an a posteriori upper bound for the error of the approximation. Since initially we are dealing with P_0, the final result consists of the approximate zeros and error bounds corresponding to P_0.

For the polynomial

$$P_8 = 7.7182 \ z^8 - 5545.9 \ z^7 - 288.08 \ z^6 + 3803.4 \ z^5 + 105.40 \ z^4$$

$$- 54.646 \ z^3 + 8191.9 \ z^2 + 8.7029 \ z - 167.00$$

considered in Section 5.1, the following output was obtained:

ZEROS		ABSOLUTE ERROR
0.7185993324953141D 03	0.0	0.19153630E-03
-0.9800090620144605D 00	-0.5479519246312825D 00	0.12113819E-03
-0.9800090620144605D 00	0.5479519246312825D 00	0.12113819E-03
-0.1433052321257051D 00	0.0	0.19153630E-03
0.1422773404427950D 00	0.0	0.19153630E-03
0.1208543646181452D 01	0.0	0.19153630E-03
0.3507680367930439D 00	0.9202200291251945D 00	0.12113819E-03
0.3507680367930439D 00	-0.9202200291251945D 00	0.12113819E-03

5. Numerical Results

5.1 Polynomials with real coefficients

The program was tested on a number of polynomials, among them four real polynomials P_N of degree $N=8,10,15,18$, taken from Ref. [2]. These polynomials either have random coefficients (P_8, P_{15}, P_{18}) or were constructed from assumed zeros (P_{10}). Calculations were in long precision. The following strategies were adopted:

(A) T_2 used exclusively.

(B) $T_1(K_N)$ is used initially. If at any step $h-1$ the number of suspect squares exceeds 250 (if $h \leq 20$) or 125 (if $h > 20$), then T_2 is used at step h. If T_2 has been used at least once and the number of suspect squares does not exceed the above limits, T_3 (modified as indicated in Section 3.4) is used.

(C) T_4 used exclusively.

(D) T_4 is used until $h = 10$, then alternatively T_3 and T_2 as in strategy (B).

Figures 3a and 3b show the number of suspect squares, NQ_h, as a function of h for the polynomial P_8 and for each of the four strategies.

The computing time for strategy (D) was found to be less than half of the computing time for all other strategies.

5.2 Polynomials with complex random coefficients

Polynomials of degree $N = 10, 20, 30$ have been constructed by using random digits. The mantissa of each coefficient was formed of six decimal digits taken from the first columns of Ref. [5]. The sign and the exponent (base 10) are determined in a similar manner. The range of the exponent was fixed to be $[-4,3]$ for $N = 10$, $[-2,2]$ for $N = 20$ and $[-1,1]$ for $N = 30$. By accident the polynomial of degree $N = 10$ presents a case of clustering of zeros around the origin: nine zeros lie inside a circle of radius <2, and the remaining zero has absolute value of about 5100.

5.3 Some special polynomials

In order to test the capability of the program to separate close zeros we applied it to the polynomials

$$P(z) = (z - \tfrac{1}{2})^2 + \varepsilon,$$

where $\varepsilon = 2^{-m}$, $m = 1, 2, \ldots$. The two complex conjugate zeros were separated up to $m = 45$. The transition from two approximate zeros to one approximate zero of multiplicity 2 is shown in Fig. 4.

To test the ability of the program to cope with many equimodular zeros, we considered the polynomials

$$P(z) = z^m + 1$$

for $m = 4, 12, 36$. In each of these cases, all zeros were determined with an error $<0.2 \times 10^{-6}$. Figure 5 shows the set S_h for $P(z) = z^{36} + 1$, h running from 2 to 7.

$P(z) = 7.7182z^8 - 5545.9z^7 - 288.08z^6 + 3803.4z^5 + 105.40z^4 - 54.646z^3 + 8191.9z^2 + 8.7029z - 167.00$

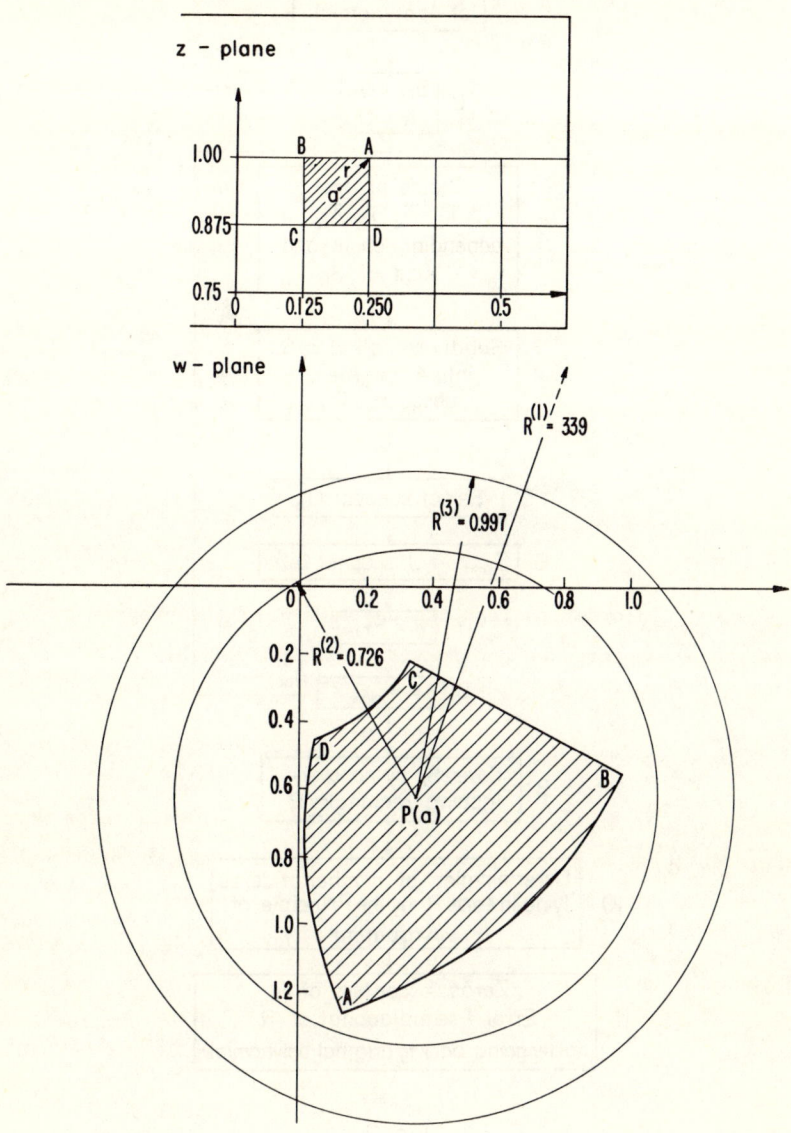

Fig. 1

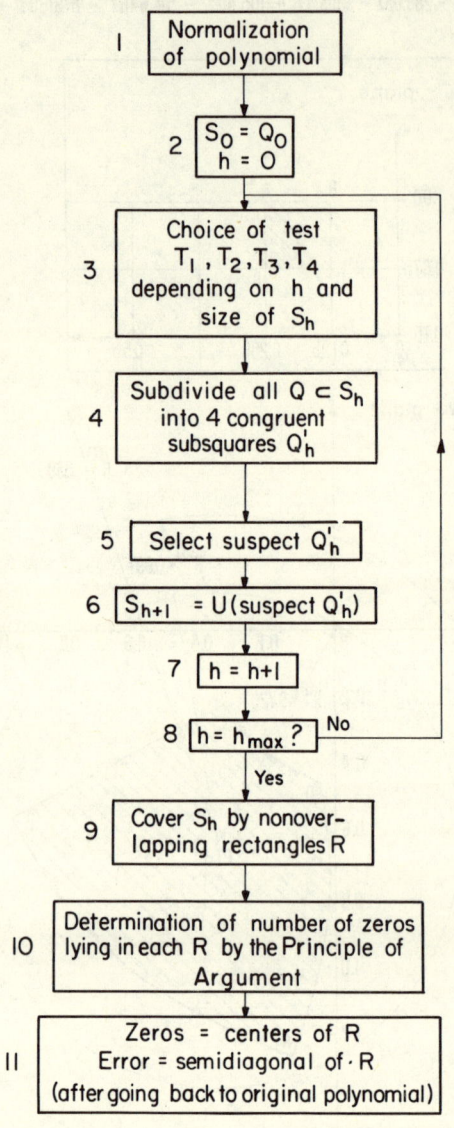

Fig. 2

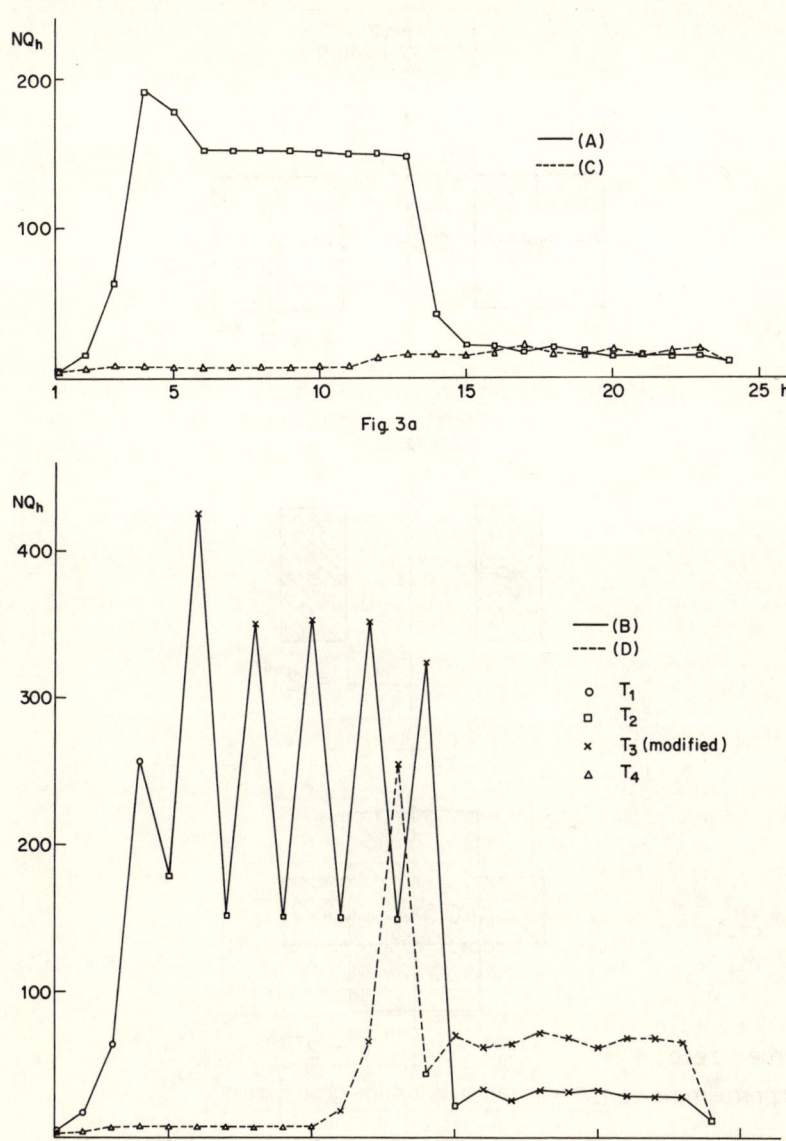

Fig. 3a

Fig. 3b

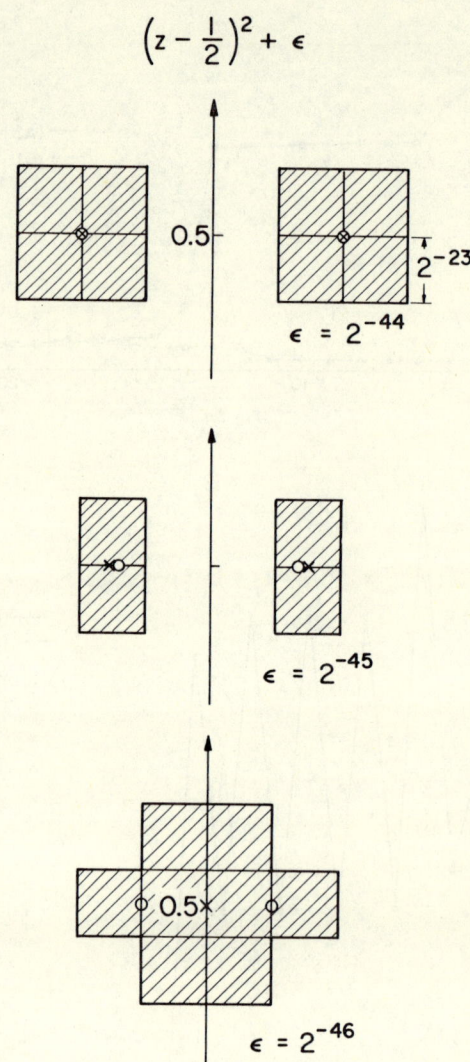

Fig. 4

$$z^{36} + 1 = 0$$

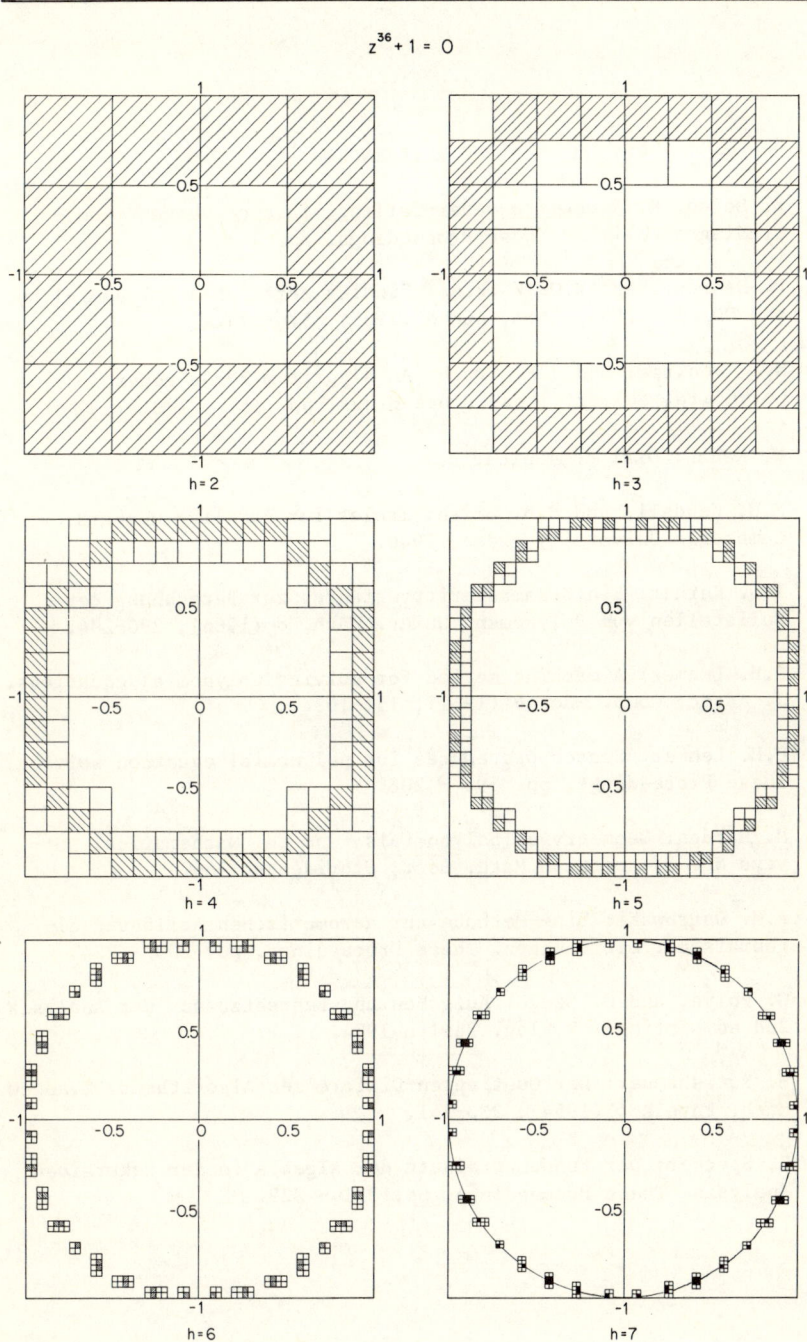

REFERENCES

1. B. Dejon, K. Nickel: A never-failing, fast-convergent root-finding algorithm. These Proceedings, pp. 1 - 35.

2. P. Henrici, and B.O. Watkins: Finding zeros of a polynomial by the QD algorithm. Comm. ACM *8* (1965), 570-574.

3. M.A. Jenkins, and J.F. Traub: An algorithm for a general polynomial solver. These Proceedings, pp. 151 - 180.

4. W. Kahan: Oral communication.

5. M.G. Kendall, and B.B. Smith: Tables for sampling numbers. Cambrdige University Press, 1946.

6. I.O. Kerner: Ein Gesamtschrittverfahren zur Berechnung der Nullstellen von Polynomen. Numer. Math. *8* (1966), 290-294.

7. D.H. Lehmer: A machine method for solving polynomial equations. J. Assoc. Comp. Mach *8* (1961), 151-162.

8. D.H. Lehmer: Search procedures for polynomial equation solving. These Proceedings, pp. 193 - 208.

9. M. Marden: Geometry of polynomials, 2nd ed. Mathematical surveys No. 3, American Math. Soc., Providence 1966.

10. A.M. Ostrowski: Eine Methode zur automatischen Auflösung algebraischer Gleichungen. These Proceedings, pp. 209 - 224.

11. G. Polya, and G. Szegö: Aufgaben und Lehrsätze aus der Analysis. 2nd ed. Springer-Verlag, Berlin 1954.

12. H. Rutishauser: Der Quotienten-Differenzen-Algorithmus. Z.angew. Math. Physik *5* (1954), 233-251.

13. E. Specker: Der Fundamentalsatz der Algebra in der rekursiven Analysis. These Proceedings, pp. 321 - 329.

14. K. Weierstrass: 1. Neuer Beweis des Fundamentalsatzes der Algebra. Ges. Werke *1*, 247-256. 2. Neuer Beweis des Satzes, dass jede ganze rationale Funktion einer Veränderlichen dargestellt werden kann als Produkt aus linearen Funktionen derselben Veränderlichen. Ges. Werke *3*, 251-269.

15. H. Weyl: Randbemerkungen zu Hauptproblemen der Mathematik, II. Fundamentalsatz der Algebra und Grundlagen der Mathematik. Math. Z. *20* (1924), 131-150.

Prof. P. Henrici
Eidgenössische Technische Hochschule
CH-8006 Zurich
Switzerland

Dr. I. Gargantini
IBM Zurich Research Laboratory
CH-8803 Rüschlikon
Switzerland

H. Hermes

On the Notion of Constructivity

0. Introduction

In this paper, which is intended as an introduction to the symposium on Constructive Aspects of the Fundamental Theorem of Algebra, I should like to present some remarks concerning the notion of constructivity. The essential concepts belonging to this topic have been formulated about thirty years ago. Thus, the concept of *recursivity* was created and it was possible on this basis to show that some interesting problems are devoid of a constructive solution.

Looking at the titles of the papers announced for this meeting I have found that the main interest consists in giving positive results concerning constructions, and not proofs for unsolvability. Hence I do not want to stress too much the negative results of the theory of recursive functions.

After having shown that a certain problem has a constructive solution, i.e. that a certain function is computable, determining the "simplicity" of a solution becomes a main interest. Such simplicity may be measured by reference to a hierarchy in the class of recursive functions. Here I should like to mention a few points concerning hierarchies of primitive recursive functions. Finally, the notion of primitive recursivity is investigated.

1. The Domain and the Range of Computable Functions

To construct means to arrange or to rearrange certain objects. It is therefore required that these objects can be manipulated. The domain and the range of a computable function should consist of manipulable (*handhabbar*) objects.

The simplest (potentially) infinite domain of manipulable objects is the domain of the natural numbers given in the form of strings of strokes. An n-tuple of natural numbers may be considered as a single word (like $||,||||,|||$). Hence it is manipulable. In particular the integers and the rationals are manipulable since they can be represented by pairs or by triples of natural numbers, respectively. (The fact that the representation is not unique — e.g. the triples a,b,c and A,B,C represent the same rational number if and only if $Ca + cB = cA + Cb$ and $cC \neq 0$ — does not alter the situation.)

A problem arises concerning the real (and the complex) numbers. The domain R of real numbers in the classical sense of this concept is undenumerable. It seems reasonable to postulate that there is only a denumerable subset of R whose elements may be considered manipulable. It is suggested that a real number is manipulable if and only if it is computable. A computable real number is given by the instructions for its computation, which are finite and manipulable. More precisely, a real number can be given by a convergent sequence $\varphi(n)$ of rational numbers. A computable real number should be given by a computable sequence $\varphi(n)$. It is desirable to require in addition that the convergence itself be computable, i.e. that there is a computable function $\psi(n)$ (whose values are natural numbers) such that $|\varphi(n_1) - \varphi(n_2)| < 1/n$ for all n_1, n_2 with $n_1, n_2 \geq \psi(n)$. Two computably convergent computable sequences

φ_1 and φ_2 for which there is a computable function χ such that $|\varphi_1(n) - \varphi_2(n)| < 1/k$ for each $n \geq \chi(k)$ are considered as equivalent.

Based on these notions an interesting theory of constructive analysis has been developed by Goodstein (cf. Goodstein [1]).

2. A Precise Definition of the Concept of a Computable Function

For every domain of manipulable objects there should be at least one effective 1 - 1 mapping* on the natural numbers (arithmetization, Gödelization). Hence, the natural numbers may be considered as a prototype for every domain of manipulable objects. It is therefore sufficient to give a precise meaning to the notion of a computable function which is defined for every natural number and whose values are natural numbers. The μ-recursive functions are of this kind. A function is called μ-*recursive* if and only if it may be obtained, starting with the *initial functions*

C_0^0 (0-place function with value 0),

S^1 (1-place function whose value is the successor of the argument),

U_i^n (for each n and each i with $1 \leq n$ and $1 \leq i \leq n$), (n-place function whose value is the i'th argument),

by a finite number of applications of the following three operations:

(i) *Substitution*, which leads from g (r-place) and $h_1,\ldots,h_r$ (all n-place) to the n-place function f for which

---

* A mapping of this kind is called *effective* if there is a procedure to compute the value for each argument and the argument for each value.

$$f(x_1,\ldots,x_n) = g(h_1(x_1,\ldots,x_n),\ldots,h_r(x_1,\ldots,x_n)),$$

(ii) *Elementary inductive definition*, which leads from g (n-place) and h ($n+2$-place) to the $n+1$-place function f for which

$$f(x_1,\ldots,x_n,0) = g(x_1,\ldots,x_n)$$

$$f(x_1,\ldots,x_n,S(y)) = h(x_1,\ldots,x_n,y,f(x_1,\ldots,x_n,y)).$$

(iii) *Application of the μ-operator to a regular function*, which leads from g ($n+1$-place) (which is supposed to be regular, i.e. that for all $x_1,\ldots,x_n$ there is at least one y such that $g(x_1,\ldots,x_n,y) = 0$) to the n-place function f for which

$$f(x_1,\ldots,x_n) = \mu y g(x_1,\ldots,x_n,y) = 0$$

where "μy" means "the smallest number y such that".

It is obvious that every μ-recursive function is computable. On the other hand the experience up to today has shown that every function which merits being called computable has the property of being μ-recursive. Church's thesis (1936) states that every computable function is μ-recursive. (cf. Church [1].)

There are other precise notions such as the concepts of a *recursive function* or a *Turing-computable function* which have been shown to be equivalent to the notion of a computable function. These concepts are investigated in the important mathematical theory of recursive functions which has been inaugurated among others by Church, Gödel, Kleene, Post and Turing. (cf. e.g. Kleene [1] and Hermes [1].)

3. An Example for the Application of Church's Thesis

In the following a proof is sketched for the well-known (cf. Goodstein [1]) fact that it is not decidable whether any two given computably convergent computable sequences of rational numbers are equivalent (cf. no. 1). The proof is indirect. *Let us assume that the equivalence relation just mentioned be decidable.* Then we proceed by the following steps:

(1) It is (under our assumption) decidable whether a 1-place recursive function F, with $F(n) \leq F(n+1)$ for all n, is bounded. P r o o f : The sequence $f(n) = 1/1 + F(n)$ is a computably convergent computable sequence of rational numbers (to show the computable convergence consider the cases where F is bounded or not). F is unbounded if and only if the sequence $f(n)$ is equivalent to the sequence $g(n)$ where $g(n)$ is identically 0.

(2) It is decidable whether a 2-place computable function G is regular (cf. no. 2). P r o o f : We define functions H and F by

$$\begin{cases} H(x,0) = 0 \\ H(x,y+1) = \overline{sg}(G(x,y) \cdot \overline{sg}(H(x,y))), \end{cases}$$

$$\begin{cases} F(0) = 0 \\ F(y+1) = F(y) + H(F(y),y), \end{cases}$$

where $\overline{sg}$ is a function with $\overline{sg}(0) = 1$ and otherwise $\overline{sg}(n) = 0$. H and F are computable. Obviously $F(y) \leq F(y+1)$. G is regular if and only if F is unbounded: (a) Let G not be regular. Then there is a number x_0 such that $G(x_0,y)$ is never 0. This implies $H(x_0,y) = 0$ for all y. We want to show that $F(y) \leq x_0$ for all y. If $F(y) < x_0$ then $F(y+1) \leq x_0$ since H has only the values 0 or 1. If $F(y) = x_0$ then

$H(F(y),y) = H(x_0,y) = 0$; hence $F(y+1) = F(y) = x_0$. (b) Let F be bounded. Then there is a number y_0 and a number x_0 such that $F(y) = x_0$ for each $y \geq y_0$. Let $y \geq y_0$. From the definition of F we deduce that $H(F(y),y)) = 0$. Hence $H(x_0,y) = 0$ for each $y \geq y_0$. Now we obtain $H(x_0,y) = 0$ for every y (because $H(x_0,y) = 1$ implies $H(x_0,y+1) = 1$). This leads to the conclusion that $G(x_0,y)$ is never 0.

(3) There is a 2-place computable function $A(k,x)$ such that for every 1-place recursive function $E(x)$ there is a number k with $E(x) = A(k,x)$ for each x. P r o o f : According to Kleene's normal form theorem (cf. for example Kleene [1] or Hermes [1]) there is a 1-place recursive function U and a 3-place recursive function T such that for each 1-place recursive function E there is a number t_E for which we have:

a) the 2-place function $Tt_E xy$ is regular,
b) $E(x) = U(\mu y\ Tt_E xy=0)$ for every x.

For each $t = 0,1,2,\ldots$ we can (according to (2)) decide whether the 2-place function $Ttxy$ is regular. Let be $t_0\ t_1\ t_2 \ldots$ the sequence of t's such that the 2-place function $Ttxy$ is regular. Let

$$A(k,x) = U(\mu y\ Tt_k xy = 0),$$

then A has the required properties.

(4) Let be $B(x) = A(x,x) + 1$. B is computable.

(5) B is not recursive. P r o o f: Otherwise according to (3) there would be a number k such that $B(x) = A(k,x)$ for every x. For $x = k$ we get a contradiction.

Now the indirect proof is finished, because (4) and (5) are contrary to Church's thesis.

4. Hierarchies of the Class of Primitive Recursive Functions

To every r-place relation R between natural numbers corresponds its *characteristic function* f_R, which is defined by

$$f_R(x_1,\ldots,x_r) = 0 \qquad iff \quad Rx_1\ldots x_r$$

$$f_R(x_1,\ldots,x_r) = 1 \qquad \text{otherwise.}$$

Hence it is possible to measure the simplicity of R by the simplicity of f_R. There is a remarkable subclass of the class of all recursive functions which is called the class of *primitive recursive functions*. A function is called primitive recursive if and only if it may be obtained starting with the initial functions (cf. no. 2) by a finite number of substitutions and of elementary inductive definitions. A primitive recursive function can be considered as more simple than a recursive function whose definition requires a μ-operator.

$\mathbf{H}_n (n=0,1,2,\ldots)$ is called a *hierarchy* of the class of primitive recursive functions, if and only if the following conditions are satisfied:

(a) each element of $\mathbf{H}_n$ is a primitive recursive function,
(b) for each n, $\mathbf{H}_n$ is a proper subclass of $\mathbf{H}_{n+1}$,
(c) for every primitive recursive function f there is a number n such that $f \varepsilon \mathbf{H}_n$.

The earliest known hierarchy $\mathbf{E}_n$ of the class of primitive recursive functions is due to Grzegorczyk [1]). The 2-place functions F_n are defined by

$$F_0(x,y) = y+1,$$
$$F_1(x,y) = x+y,$$
$$F_2(x,y) = (x+1)(y+1),$$
for $n \geq 2$:
$$\begin{cases} F_{n+1}(0,y) = F_n(y+1,y+1) \\ F_{n+1}(x+1,y) = F_{n+1}(x,F_{n+1}(x,y)). \end{cases}$$

A function belongs to $\mathbf{E}_n$ if and only if it may be obtained starting with the initial functions and F_n by a finite number of substitutions and restricted elementary inductive definitions. The process of *restricted elementary inductive definition* leads from g (n-place), h ($n+2$-place) and k ($n+1$-place) to the $n+1$-place function f if and only if the three following conditions are satisfied:

(1) $\qquad f(x_1,\ldots,x_n,0) = g(x_1,\ldots,x_n)$

(2) $\qquad f(x_1,\ldots,x_n,y+1) = h(x_1,\ldots,x_n,y,f(x_1,\ldots,x_n,y))$

(3) $\qquad f(x_1,\ldots,x_n,y) \leq k(x_1,\ldots,x_n,y).$

$\mathbf{E}_3$ coincides with the class of *elementary functions*. A function is called elementary if and only if it may be obtained starting with the initial functions, and the sum and the modified difference* by a finite number of substitutions and applications of the operators $\Sigma_{\leq y}$ and $\Pi_{\leq y}$ where

$$(\Sigma_{\leq y} f)(x_1,\ldots,x_n) = \sum_{z \leq y} f(x_1,\ldots,x_n,z)$$

and $\qquad (\Pi_{\leq y} f)(x_1,\ldots,x_n) = \prod_{z \leq y} f(x_1,\ldots,x_n,z).$

---
* the modified difference of x and y is $x-y$ if $x \geq y$, and 0 otherwise.

Another hierarchy $\mathbf{A}_n$ of the class of primitive recursive functions has been given by Axt [1] and independently by Heinermann [1]. A function belongs to $\mathbf{A}_n$ if and only if it may be obtained starting with the initial functions by a finite number of substitutions and at most n elementary inductive definitions. For $n \geq n_0$ we have $\mathbf{A}_n = \mathbf{E}_{n+1}$ (cf. e.g. Meyer [1]).

5. Primitive Recursive Functions Versus Recursive Functions

If we believe in Church's thesis (and most mathematicians do) we have to admit that the class of recursive functions is not merely a useful mathematical concept (like the class of continuous real functions) but has in addition a fundamental significance. One may ask the same question for the class of *primitive* recursive functions. The definition of primitive recursiveness given in no. 4 was obtained from the definition of recursiveness by omitting one of the generating processes. This looks rather artificial. It does not seem possible to give an absolute meaning to the definition of primitive recursivity by defining this notion using only the concept of recursivity.

Sometimes the idea is ventured that a primitive recursive function is distinguished among the recursive functions by the fact that the number of steps which are required to its computation for a given argument may be more easily estimated.

This is obvious, since - if $S_f(x_1,\ldots,x_n)$ is the number of steps for the computation of $f(x_1,\ldots,x_n)$ e.g. by a given Turing-machine - S_f is primitive recursive, if f is primitive recursive. But it is also the case, that S_f is recursive, if f is recursive. Hence this idea is unlikely to lead to an "absolute" definition of primitive recursivity.

There are facts which indicate a certain artificiality of the concept of primitive recursiveness. E. Specker [1] has proved that there are primitive recursive real numbers* without a primitive recursive decimal.

Nevertheless it is undeniable that primitive recursivity is a very useful concept. Nearly every function occurring in practice whose domain and range is the class of natural numbers is primitive recursive. Kleene's normal form theorem (cf. no. 3 for the special case of binary functions; U and T could be chosen as primitive recursive) shows that there is a very simple way leading from primitive recursive functions to arbitrary recursive functions.

6. The Definition of Primitive Recursivity for the Domain of Natural Numbers and for Other Domains

In no. 5 we have remarked that the definition of primitive recursive function was obtained from the definition of recursive function by omitting one of the generating processes. Now let us look at the remaining processes. Among them we include the initial functions.

We can distinguish two kinds of generating processes. Some depend on the nature of the domain (which is here the domain of natural numbers), others not. The functions U_i^n and the substitution are independent of the domain. On the other hand the processes C_0^0, N^1 and the elementary inductive definition refer to the structure of the domain. The domain of natural numbers can be con-

---

* i.e. real numbers given by the method of no. 1, where the occurring computable functions are primitive recursive.

sidered as an absolute free algebra of the type $<0,1>$*. C_0^0 and N^1 refer to the generating functions of this algebra and the process of elementary inductive definition operates according to these functions.

The notion of primitive recursivity, as defined in no. 4, is confined to the domain **N** of natural numbers. In order to extend this concept to other infinite domains **D** of manipulable objects two procedures may be considered:

(1) It is possible to transfer the class **P** of ordinary primitive recursive functions by means of a Gödelization γ of **D** (cf. no. 2). Let $X_1,\ldots,X_n$ be variables for elements of **D**. Then we transfer each function f over **N** into a function f_γ over **D** by the definition

$$f_\gamma(X_1,\ldots,X_n) = \gamma^{-1}(f(\gamma(X_1),\ldots,\gamma(X_n))).$$

Let $\mathbf{P}_\gamma$ be the class of all f_γ with $f \varepsilon \mathbf{P}$. The elements of $\mathbf{P}_\gamma$ may be called γ-*primitive recursive functions*.

(2) If **D** is endowed with a structure **S** (like **N**) it may be possible to imitate the definition of ordinary primitive recursive functions by choosing (besides the U_i^n and the substitution) processes which depend on **S**. We thus get the class $\mathbf{P_S}$ of *genuine primitive recursive functions over* **D(S)**.

Both procedures are not without problems. In general different Gödelizations γ_1, γ_2 lead to different classes $\mathbf{P}_{\gamma_1}, \mathbf{P}_{\gamma_2}$ which

---

* The elements of an absolute free algebra **A** of the type $<n_1,\ldots,n_r>$ are uniquely generated by n_j-place functions f_j ($j=1,\ldots,r$). A 0-place function is considered as an element of **A**.

makes (1) ambiguous. Concerning (2) it may be remarked: (a) It is not clear whether a domain **D** of manipulable objects has a structure **S** at all. (b) It could be that **D** can be considered as endowed with different structures S_1, S_2 which may lead to different classes P_{S_1}, P_{S_2}. (c) The imitation of the definition of the ordinary primitive recursive functions may be problematic for **D(S)**.

I want to say that we have *a satisfactory solution for* **D** if the following two conditions are fulfilled:

(i) **D** is endowed with a "canonical" structure **S** and there is a natural way to imitate the ordinary definition of primitive recursiveness in order to get the class P_S of genuine primitive recursive functions.

(ii) There is a Gödelization γ such that $P_S = P_\gamma$.

7. Examples

I want to give some examples for the problem discussed in the preceding number.

(1) The class **D** of words over a given alphabet with m elements $a_1,\ldots,a_m$ (including the empty word E) may be considered in a canonical way as an absolute free algebra of the type $<0,1,\ldots,1>$, where the 0-place function is E and the different 1-place functions f_j concatenate their argument with the different letters a_j. We are able to imitate the definition of the ordinary primitive recursive functions by choosing (besides the U_i^n and the substitution) the following generating processes:

(a) (corresponding to C_0^0) the 0-place function C_E^0, which has as value the empty word,

(b) (corresponding to N^1) the 1-place functions f_j,

(c) (corresponding to the elementary inductive definition) the *generalized elementary inductive definition* which leads from g (n-place) and the functions $h_1,\ldots,h_m$ ($n+2$-place) to the $n+1$-place function f for which

$$f(X_1,\ldots,X_n,E) = g(X_1,\ldots,X_n)$$

$$f(X_1,\ldots,X_n,f_j(Y)) = h_j(X_1,\ldots,X_n,Y,f(X_1,\ldots,X_n,Y)) \qquad (j=1,\ldots m)$$

Asser [1] has shown that there is a Gödelization γ such that $P_s = P_\gamma$.

Mahn [1] has generalized the result of Asser for the case of an absolute free algebra of *arbitrary finite type*.

Péter [1] considers absolute free algebras of an arbitrary type which are endowed with a "predecessor-relation". Using this relation she gives a definition of primitive recursive functions without investigating whether these functions can be obtained by transferring the ordinary primitive recursive functions by means of an appropriate Gödelization.

(2) The following examples deal with algebras which are generated by a finite number of operations, but with elements without a unique representation by the generating functions. Even in this case it is possible to generalize the process of elementary inductive definition. The occurring equations may be contradictory. Hence we have to restrict ourselves to the case of non-contradiction.

(2a) The domain **D** of ordered pairs of natural numbers may be considered in a canonical way as an algebra of the type $\langle 0,1,1\rangle$ where the 0-place function is the pair $\langle 0,0\rangle$ and the two 1-place functions add the number 1 to the first resp. to the second component of the argument. It is easy to show that we have a satisfactory solution in this case.

(2b) The domain **D** of *totally finite sets* may be defined as the smallest class of sets to which belongs the empty set and which is closed under the 1-place operation { } (unit set) and the 2-place operation ∪ (union). Hence **D** may be considered as an algebra of type $\langle 0,1,2\rangle$. Rödding [1] has proved that we have a satisfactory solution in this case.

(2c) The domain **D** of words (without the empty word) over a finite alphabet may be considered in a canonical way as an algebra of type $\langle 0,\ldots,0,2\rangle$ where the 0-place functions are the letters of the alphabet and the 2-place function is the concatenation. It is not known whether this problem has a satisfactory solution, even in the case of a 1-element alphabet.

REFERENCES

Asser, G. [1] Rekursive Wortfunktionen. Zeitschrift für math. Logik und Grundlagen der Math. *6* (1960) 258-278.

Axt, P. [1] Iteration of primitive recursion. Zeitschrift für math. Logik und Grundlagen der Math. *11* (1965), 253-255.

Church, A. [1] An unsolvable problem of elementary number theory. American Journal of Math. *58* (1936) 345-363.

Goodstein, R.L. [1] Recursive analysis. North Holland Publishing Company, Amsterdam 1961.

Grzegorczyk, A. [1] Some classes of recursive functions. Rozprawy
    Matematyczne *4* (1953) 1-45.

Heinermann, W. [1] Untersuchungen über die Rekursionszahlen rekursiver Funktionen. Diss. Münster 1961. Unpublished.

Hermes, H. [1] Enumerability, decidability, computability. Springer,
    Berlin/Heidelberg/New York 1965.

Kleene, S.C. [1] Introduction to metamathematics. North Holland
    Publishing Company, Amsterdam 1952.

Mahn, F.K. [1] Ueber die Strukturabhängigkeit des Begriffs der
    primitiv-rekursiven Funktion. To appear in Archiv für
    math. Logik und Grundlagenforschung *12*.

Meyer, A.R. [1] Depth of nesting and the Grzegorczyk hierarchy.
    Abstract Amer. Math. Soc. Notices *12* (1965) 342.

Péter, R. [1] Ueber die Verallgemeinerung der Theorie der rekursiven Funktionen für abstrakte Mengen geeigneter Struktur
    als Definitionsbereiche. Acta Math.Hung. *12* (1961)
    271-314.

Rödding, D. [1] Primitiv-rekursive Funktionen über einem Bereich
    endlicher Mengen. Archiv für math. Logik und Grundlagenforschung *10* (1967) 13-29.

Specker, E. [1] Nicht konstruktiv beweisbare Sätze der Analysis.
    J. symbolic Logic *14* (1949) 145-158.

Prof. Dr. H. Hermes
Mathematisches Institut
Albert-Ludwigs-Universität
D-78 Freiburg i.Br.

A. S. Householder and G. W. Stewart, III

Bigradients, Hankel Determinants, and the Padé Table*

Introduction

This note might perhaps be entitled "An essay on bigradients and related topics", since it is an effort to throw possibly new light onto these matters and to provide a unification that seems to be lacking in the existing literature. Only Theorem 2 and the Lemma by which it is proved are possibly new, although special cases of the theorem are found in many places. Theorem 1 goes back more than a century to Trudi, but seems to have been overlooked almost completely. Theorem 3, which provides an explicit representation for the entries in the Padé table, was given by Frobenius, but is seldom if ever exploited in the current literature (however, see [8]). From these three theorems, along with a classical determinantal identity, can be obtained quite directly much of the algebraic theory (as distinguished from theorems on convergence) of the qd algorithm, the counting of zeros in a disk or half-plane, criteria for representing a power series or a rational fraction as a continued fraction of specified type, and other matters.

---

*) Research sponsored by the U.S. Atomic Energy Commission under contract with the Union Carbide Corporation.

1. Bigradients

A bigradient is a determinant formed in a certain way from two sequences, possibly infinite. If the sequences are

(1.1)
$$a: a_0, a_1, a_2, \ldots$$
$$b: b_0, b_1, b_2, \ldots,$$

then the bigradient (i,j) is

(2)
$$\delta \begin{Bmatrix} (a)_i \\ (b)_i \end{Bmatrix} = \delta \begin{pmatrix} a_0 & a_1 & \cdots & a_{i+j-1} \\ 0 & a_0 & \cdots & a_{i+j-2} \\ \cdot & \cdot & \cdot & \cdot \\ 0 & b_0 & \cdots & b_{i+j-2} \\ b_0 & b_1 & \cdots & b_{i+j-1} \end{pmatrix}$$

where the δ singifies the determinant. It will be understood that if either sequence is finite, zeros are to be adjoined to make it infinite.

In case the sequences are sequences of coefficients of polynomials,

(1.3)
$$f(z) = a_0 z^n + a_1 z^{n-1} + \ldots + a_n,$$
$$g(z) = b_0 z^m + b_1 z^{m-1} + \ldots + b_m,$$

the polynomial bigradient (i,j) will be defined by

(1.4) $\delta \begin{pmatrix} (f)_i \\ (g)_j \end{pmatrix} = \delta \begin{pmatrix} a_0 & a_1 & \cdots & a_{i+j-2} & z^{i-1}f \\ 0 & a_0 & \cdots & a_{i+j-3} & z^{i-2}f \\ \cdots & \cdots & \cdots & \cdots & \cdots \\ 0 & b_0 & \cdots & b_{i+j-3} & z^{j-2}g \\ b_0 & b_1 & \cdots & b_{i+j-2} & z^{j-1}g \end{pmatrix}$

In this case the bigradients for which $i+n = j+m$ are of particular interest:

Lemma 1

If f and g are given by (3), then

(1.5) $\delta \begin{pmatrix} (f)_{m-j} \\ (g)_{n-j} \end{pmatrix} = z^j \delta \begin{pmatrix} (a)_{m-j} \\ (b)_{n-j} \end{pmatrix} + \cdots$

where the terms omitted are of degree $j-1$ and less in z.

This is verified by an obvious determinantal reduction.

Now each polynomial bigradient is of the form

(1.6) $\delta \begin{pmatrix} (f)_i \\ (g)_j \end{pmatrix} = \varphi_i f + \psi_j g,$

where φ_i is of degree $i-1$ at most and ψ_j of degree $j-1$ at most and this fact can be used to establish

Theorem 1 (Trudi)

For the polynomials f and g in (1.3) let

$$(1.7) \quad 0 = \delta\begin{pmatrix} (a)_m \\ (b)_n \end{pmatrix} = \delta\begin{pmatrix} (a)_{m-1} \\ (b)_{n-1} \end{pmatrix} = \ldots = \delta\begin{pmatrix} (a)_{m-j+1} \\ (b)_{n-j+1} \end{pmatrix} \neq \delta\begin{pmatrix} (a)_{m-j} \\ (b)_{n-j} \end{pmatrix}.$$

Then f and g have a common divisor of degree j, and it is given by the right member of (1.5). Conversely, if f and g have a common divisor of degree j but not higher, then (1.7) are satisfied, and again the divisor is given by the right member of (1.5).

For $j=0$, (1.5) gives the ordinary resultant, and Trudi's theorem includes as a special case the usual theorem about the resultant. The complete theorem can be proved by a kind of induction. To begin with, note that the left member of (1.5), hence also the right member, vanishes whenever z is set equal to a common zero for f and g. In particular, if they have at least one, then the right member of (1.5) for $j = 0$ must equal zero. Conversely, if the resultant vanishes, then the combination (1.6) with $i = m$ and $j = n$ vanishes identically. But φ_m is of degree only $m-1$, hence cannot contain all linear factors of g, hence f and g have in common at least one linear factor.

When this is true, then (1.5) with $j = 1$ either vanishes identically, or is equal to the common factor, since when this common factor vanishes both sides vanish. But if the leading coefficient on the right is equal to zero, so also is the constant term as can be seen by setting z equal to the zero common to f and g. Then f and g have at least a quadratic factor in common. Clearly the argument can be continued.

This and related results by Trudi are summarized by Muir [6]; also the theorem is given in part by Bôcher [2], who, however, fails to exhibit the form (1.5) for the common divisor. More recently Bareiss [1] has made use of a rather special case of the theorem in a certain algorithm for solving an algebraic equation.

2. Hankel Determinants

Next, consider the following power series:

$$f(z) = a_0 + a_1 z + a_2 z^2 + \ldots ,$$
(2.1)
$$g(z) = b_0 + b_1 z + b_2 z^2 + \ldots ,$$
$$h(z) = c_0 + c_1 z + c_2 z^2 + \ldots ,$$

and let

(2.2) $$h(z) = g(z)/f(z) .$$

The Hankel determinants for h are defined by

(2.3) $$H_\nu^{(p)} = H_\nu^{(p)}[c] = \delta \begin{pmatrix} c_\nu & c_{\nu+1} & \cdots & c_{\nu+p-1} \\ c_{\nu+1} & c_{\nu+2} & \cdots & c_{\nu+p} \\ \cdots & \cdots & \cdots & \cdots \\ c_{\nu+p-1} & c_{\nu+p} & \cdots & c_{\nu+2p-2} \end{pmatrix} .$$

Understanding that

$$0 = c_{-1} = c_{-2} = \ldots ,$$

these are defined for all indices ν, and with the convention

$$H_\nu^{(0)} = 1,$$

for all $p \geq 0$.

Theorem 2

If f, g, and h are as given in (2.1) and satisfy (2.2) then

$$(2.4) \qquad \delta \begin{pmatrix} (a)_\nu \\ (b)_p \end{pmatrix} = a_0^{\nu+p} \, H_{\nu-p+1}^{(p)} \, [c] .$$

A special case of this is the set of Wronski identities on which a progressive form of the qd algorithm is based.

The proof of this will be based upon the following lemma, which is in itself of some interest:

Lemma 2

Let A be a nonsingular matrix of order n, and let X and B be $n \times m$ matrices satisfying

$$AX = B .$$

Let σ represent a set of m distinct indices chosen from the set $1, 2, \ldots, n$; let X_σ represent the matrix of order m whose rows are those of X having indices in the set σ; let A_σ represent the *matrix obtained on inserting the columns of* B *in place of those in* A *whose indices belong to* σ. *Then*

$$\delta(X_\sigma) = \delta(A_\sigma)/\delta(A) .$$

The proof requires only the observation that in the matrix $A^{-1}A_\sigma$, each column is either a column of the identity or a column of X.

Now since $fh = g$, a comparison of coefficients leads to equations that can be written in the form (with some redundance)

$$\begin{pmatrix} a_0 & 0 & 0 & 0 & \cdots \\ a_1 & a_0 & 0 & 0 & \cdots \\ a_2 & a_1 & a_0 & 0 & \cdots \\ a_3 & a_2 & a_1 & a_0 & \cdots \\ & & \cdots & & \end{pmatrix} \begin{pmatrix} c_0 & 0 & 0 & 0 & \cdots \\ c_1 & c_0 & 0 & 0 & \cdots \\ c_2 & c_1 & c_0 & 0 & \cdots \\ c_3 & c_2 & c_1 & c_0 & \cdots \\ & & \cdots & & \end{pmatrix} \begin{pmatrix} b_0 & 0 & 0 & 0 & \cdots \\ b_1 & b_0 & 0 & 0 & \cdots \\ b_2 & b_1 & b_0 & 0 & \cdots \\ b_3 & b_2 & b_1 & b_0 & \cdots \\ & & \cdots & & \end{pmatrix}.$$

Take A to be a leading principal submatrix of the matrix of the a's; take $n = \nu+2p-1$, $m = p$, and

$$\sigma = [\nu+p, \nu+p+1, \ldots, \nu+2p-1] ,$$

and reverse the order of the first p columns in each of the other two matrices. The result comes out directly except that the determinants are in transposed form.

Corollary 1

If $g(z) = 1 = f(z) h(z)$, then

$$H^{(p)}_{\nu-p+1} [c] = (-1)^{(\nu+p)(\nu+p-1)/2} H^{(\nu)}_{p-\nu+1} [a] .$$

This is the special case required for the progressive form of the qd algorithm. It is given by Borel [3], but is in fact much older.

When f and g are polynomials of specified degree, it is easy to see that all bigradients of a doubly infinite rectangular array vanish identically, hence this must be true of the Hankel determinants. It can be seen that, on the other hand, none of the bigradients can vanish in the row and column immediately bordering the vanishing array. The necessity of these conditions falls out immediately from an examination of the bigradients. To prove sufficiency it is necessary to look at the equations to be satisfied by the coefficients of f and g.

3. The Euclidean Algorithm

The polynomial bigradients (1.5) are evidently multiples of the remainders that occur in the Euclidean algorithm. Consider the somewhat special case of the following polynomials

(3.1)
$$f_0(z) = z^n + a_{01} z^{n-1} + \ldots + a_{0n} ,$$

$$f_1(z) = z^{n-1} + a_{11} z^{n-2} + \ldots + a_{1,n-1} ,$$

and suppose the Euclidean algorithm can be carried to completion in the form

(3.2)
$$f_0 = q_1 f_1 - \beta_1 f_2,$$
$$f_1 = q_2 f_2 - \beta_2 f_3,$$
$$\ldots$$
$$f_{n-2} = q_{n-1} f_{n-1} - \beta_{n-1},$$

where

(3.3)
$$f_{n-j} = z^j + a_{n-j,1} z^{j-1} + \ldots + a_{n-j,j}.$$

Thus each polynomial is monic. Let

(3.4)
$$q_j = z - \alpha_j.$$

The problem is here to express each β_j in terms of bigradients in the coefficients of f_0 and f_1. On applying (1.5) with $j = n-2$, it is at once clear that

$$\beta_1 = \delta \begin{pmatrix} (a_0)_1 \\ (a_1)_2 \end{pmatrix},$$

and, moreover,

(3.5)
$$\beta_j = \delta \begin{pmatrix} (a_{j-1})_1 \\ (a_j)_2 \end{pmatrix}.$$

Now, if the multiplications are carried out in the first of (3.2) and coefficients compared, the following matrix identity can be verified:

$$\begin{pmatrix} 1 & 0 & 0 & \alpha_2 & -1 \\ 0 & 1 & \alpha_1 & -1 & 0 \\ 0 & 0 & 1 & 0 & 0 \\ 0 & 0 & 0 & 1 & 0 \\ 0 & 0 & 0 & 0 & 1 \end{pmatrix} \begin{pmatrix} 1 & a_{01} & a_{02} & a_{03} & a_{04} \\ 0 & 1 & a_{01} & a_{02} & a_{03} \\ 0 & 0 & 1 & a_{11} & a_{12} \\ 0 & 1 & a_{11} & a_{12} & a_{13} \\ 1 & a_{11} & a_{12} & a_{13} & a_{14} \end{pmatrix}$$

$$= \begin{pmatrix} 0 & 0 & -\beta_1 & -\beta_1 a_{21} & -\beta_1 a_{22} \\ 0 & 0 & 0 & -\beta_1 & -\beta_1 a_{21} \\ 0 & 0 & 1 & a_{11} & a_{12} \\ 0 & 1 & a_{11} & a_{12} & a_{13} \\ 1 & a_{11} & a_{12} & a_{13} & a_{14} \end{pmatrix},$$

and when determinants are taken on both sides and obvious reductions made, the result is

$$\delta \begin{pmatrix} (a_0)_2 \\ (a_1)_3 \end{pmatrix} = \beta_1^2 \; \delta \begin{pmatrix} (a_1)_1 \\ (a_2)_2 \end{pmatrix} .$$

A similar type of argument shows, more generally, that

$$(3.6) \qquad \delta \begin{pmatrix} (a_0)_j \\ (a_1)_{j+1} \end{pmatrix} = \beta_1^j \; \delta \begin{pmatrix} (a_1)_{j-1} \\ (a_2)_j \end{pmatrix} .$$

The same argument can, of course, be applied to any one of the equations (3.2) with corresponding results. From these it is verified at once that

$$(3.7) \qquad \delta \begin{pmatrix} (a_0)_j \\ (a_1)_{j+1} \end{pmatrix} = \beta_1^j \beta_2^{j-1} \cdots \beta_j,$$

and hence that

$$(3.8) \qquad \beta_i = \delta \begin{pmatrix} (a_0)_{j-2} \\ (a_1)_{j-1} \end{pmatrix} \delta \begin{pmatrix} (a_0)_j \\ (a_1)_{j+1} \end{pmatrix} / \delta^2 \begin{pmatrix} (a_0)_{j-1} \\ (a_1)_j \end{pmatrix}.$$

This result is also due to Trudi and given by Muir. When the algorithm goes through, it permits the continued fraction representation

$$(3.9) \qquad f_1/f_0 = 1/\overline{z-\alpha_1} - \beta_1/\overline{z-\alpha_2} - \beta_2/\overline{z-\alpha_3} - \cdots.$$

The usual algorithm for forming a Sturm sequence is of the form

$$(3.10) \qquad \begin{aligned} \varphi_0 &= Q_1 \varphi_1 - \varphi_2, \\ \varphi_1 &= Q_2 \varphi_2 - \varphi_3, \end{aligned}$$

where none of the polynomials is necessarily monic. If

$$\varphi_0 = \gamma_0 f_0, \quad \varphi_1 = \gamma_1 f_1,$$

then the first of (3.2) can be multiplied by γ_0 and identified with the first of (3.10), from which it follows that

$$\varphi_2 = \gamma_0 \, \beta_1 \, f_2 \, ;$$

then the second of (3.2) can be multiplied by γ_0 and identified with the second of (3.10), whence

$$\varphi_3 = \gamma_1 \, \beta_2 \, f_3.$$

In general, if $\varphi_i = \gamma_i f_i$, then

(3.11)
$$\varphi_{i+2} = \gamma_i \, \beta_{i+1} \, f_{i+2},$$

$$\gamma_{i+2} = \gamma_i \, \beta_{i+1},$$

and $\gamma_2, \gamma_3, \ldots$, are evaluated recursively.

4. The Padé Table

The usual approach to this subject is by way of a series of descending powers

(4.1) $$S(z) = s_0 z^{-1} + s_1 z^{-2} + \ldots$$

and its sections

(4.2) $$S_\nu(z) = s_0 z^{-1} + \ldots + s_{\nu-1} z^{-\nu}.$$

The Padé table, as is well known, is a doubly infinite table, each of whose entries is the quotient of two polynomials of specified degrees, chosen so that when the quotient is expanded it agrees with (4.1) to as many terms as possible. Clearly if the numerator is of degree m, the denominator of degree n, there are essentially $n+m$ coefficients to be determined, hence the agreement should go as far as the term in z^{-n-m}. It turns out to be easy to write down explicitly, in determinantal form, the required polynomials. To obtain these, consider first the determinants

$$(4.3) \quad M_\nu^{(p)}(z) = \delta \begin{pmatrix} s_\nu & s_{\nu+1} & \cdots & s_{\nu+p-1} & z^{-p} S_\nu \\ s_{\nu+1} & s_{\nu+2} & \cdots & s_{\nu+p} & z^{-p+1} S_{\nu+1} \\ & & \cdots & & \\ s_{\nu+p} & s_{\nu+p+1} & \cdots & s_{\nu+2p-1} & S_{\nu+p} \end{pmatrix}$$

and

$$(4.4) \quad L_\nu^{(p)}(z) = \delta \begin{pmatrix} s_\nu & s_{\nu+1} & \cdots & s_{\nu+p-1} & z^{-p} \\ s_{\nu+1} & s_{\nu+2} & \cdots & s_{\nu+p} & z^{-p+1} \\ & & \cdots & & \\ s_{\nu+p} & s_{\nu+p+1} & \cdots & s_{\nu+2p-1} & 1 \end{pmatrix}.$$

Theorem 3 (Frobenius)

The quotient

$$(4.5) \quad M_\nu^{(p)}(z)/L_\nu^{(p)}(z) = S_{\nu+2p}(z) + \left(z^{-\nu-2p-1} \right),$$

where the parentheses represent a sum of powers of z^{-1} of degree $\nu+2p+1$ and higher.

Clearly the quotient on the left can be made into the quotient of two polynomials in z by multiplying numerator and denominator by a suitable power of z, and the resulting fraction is therefore an entry in the Padé table. To prove the theorem it is sufficient to show that the series

$$S(z) \, L_\nu^{(p)}(z) - M_\nu^{(p)}(z)$$

begins with the term in $z^{-\nu-2p-1}$. To do this, note first that in the determinant for $M_\nu^{(p)}$, its value will remain unchanged if in the last column each index is increased by any amount up to p, since this is the result of adding to the last column previous columns multiplied by powers of z. Hence if the product $S(z) \, L_\nu^{(p)}(z)$ is formed by multiplying $S(z)$ into the last column and $M_\nu^{(p)}(z)$ subtracted, the lowest power of z^{-1} in the result will be $\nu+2p+1$. This completes the proof.

These determinants, and also the Hankel determinants, satisfy certain important three-term identities, all of which are consequences of a fundamental determinantal identity as follows: If a, b, c, and d are vectors of dimension n, and M is an $n \times (n-2)$ matrix, then

(4.6) $\quad \delta(abM) \, \delta(cdM) = \delta(acM) \, \delta(bdM) - \delta(adM) \, \delta(bcM).$

This can be established by applying a Laplace expansion to the vanishing determinant of

$$\begin{pmatrix} a & b & c & d & M & 0 \\ a & b & c & d & 0 & M \end{pmatrix}.$$

For the required application the vectors c and d are taken to be certain columns of the identity matrix.

When (4.6) is applied to a Hankel determinant with the vectors and matrix M appropriately chosen, the result is the well known identity

(4.7) $\quad H_{\nu+1}^{(p-1)} H_{\nu-1}^{(p+1)} = H_{\nu-1}^{(p)} H_{\nu+1}^{(p)} - [H_\nu^{(p)}]^2 .$

It can be applied in various ways to the determinants (4.3) and (4.4), possibly bordered in a suitable way, and among the results are

(4.8) $\quad H_{\nu+2}^{(p)} K_\nu^{(p+1)}(z) = H_\nu^{(p+1)} K_{\nu+2}^{(p)}(z) - z^{-1} H_{\nu+1}^{(p+1)} K_{\nu+1}^{(p)}(z) ,$

(4.9) $\quad H_{\nu+1}^{(p)} K_\nu^{(p)}(z) = H_\nu^{(p)} K_{\nu+1}^{(p)}(z) + z^{-1} H_\nu^{(p+1)} K_{\nu+1}^{(p-1)}(z) ,$

where $K_\nu^{(p)}(z)$ represents either $M_\nu^{(p)}(z)$ or $L_\nu^{(p)}(z)$.

Now let

(4.10) $\quad p_\nu^{(p)}(z) = z^p L_\nu^{(p)}(z)/H_\nu^{(p)}, \quad r_\nu^{(p)}(z) = z^p M_\nu^{(p)}(z)/(s_0 H_\nu^{(p)}).$

Then $p_\nu^{(p)}(z)$ and $z^\nu r_\nu^{(p)}(z)$ are monic polynomials, and the preceding identities lead directly to

(4.11) $\quad p_{\nu+1}^{(p)}(z) = p_\nu^{(p)}(z) - e_\nu^{(p)} p_{\nu+1}^{(p-1)}(z) ,$

$$(4.12) \quad r^{(p)}_{\nu+1}(z) = r^{(p)}_\nu(z) - e^{(p)}_\nu r^{(p-1)}_{\nu+1}(z) ,$$

$$(4.13) \quad r^{(p+1)}_{\nu+1}(z) = z\, p^{(p)}_{\nu+1}(z) - q^{(p+1)}_\nu p^{(p)}_\nu(z) ,$$

$$(4.14) \quad r^{(p+1)}_{\nu+1}(z) = z\, r^{(p)}_{\nu+1}(z) - q^{(p+1)}_\nu r^{(p)}_\nu(z) ,$$

where

$$(4.15) \quad \begin{aligned} q^{(p)}_\nu &= H^{(p)}_{\nu+1} H^{(p-1)}_\nu \big/ H^{(p)}_\nu H^{(p-1)}_{\nu+1} \\ e^{(p)}_\nu &= H^{(p+1)}_\nu H^{(p-1)}_{\nu+1} \big/ H^{(p)}_\nu H^{(p)}_{\nu+1} \end{aligned} .$$

The verification of these results is perhaps laborious, but entirely straightforward. The q's and e's are, of course, those of the qd algorithm.

Consider the special case of $\nu = 0$. Then $p^{(p)}_0$, $p^{(p)}_1$, $r^{(p)}_0$, and $zr^{(p)}_1$ are all monic polynomials. If each of (4.11) and (4.13) is multiplied through by z, and each of (4.12) and (4.14) by s_0, the resulting identities can be interpreted as the classical recursions for the partial numerators and denominators of the continued fraction whose approximants are

$$s_0 r^{(0)}_0 / p^{(0)}_0 , \quad s_0 z r^{(0)}_1 / (z p^{(0)}_1) , \quad s_0 r^{(1)}_0 / p^{(1)}_0 , \quad s_0 z r^{(1)}_1 / (z p^{(1)}_1) , \quad \ldots$$

and the fraction is

$$(4.16) \quad F(z) = s_0 / \overline{z} - q^{(1)}_0 / \overline{1} - e^{(1)}_0 / \overline{z} - q^{(2)}_0 / \overline{1} - \ldots$$

Naturally a similar discussion can be made for arbitrary ν, and the continued fraction, when it exists, represents $S(z) - S_\nu(z)$.

Now apply (4.11) and (4.12) twice each with indices p and $p-1$. The result is

$$(4.17) \quad p_\nu^{(p)}(z) = (z - \alpha_\nu^{(p)}) p_\nu^{(p-1)}(z) - \beta_\nu^{(p-1)} p_\nu^{(p-2)}(z),$$

$$(4.18) \quad r_\nu^{(p)}(z) = (z - \alpha_\nu^{(p)}) r_\nu^{(p-1)}(z) - \beta_\nu^{(p-1)} r_\nu^{(p-2)}(z),$$

where

$$(4.19) \quad \alpha_\nu^{(p)} = q_{\nu-1}^{(p)} + e_{\nu-1}^{(p)}, \quad \beta_\nu^{(p)} = q_{\nu-1}^{(p+1)} e_{\nu-1}^{(p)}$$

By application of (4.7) it is shown that

$$(4.20) \quad \alpha_\nu^{(p)} = q_\nu^{(p)} + e_\nu^{(p-1)}, \quad \beta_\nu^{(p)} = q_\nu^{(p)} e_\nu^{(p)}$$

and these, indeed, provide the qd algorithm. But (4.17) and (4.18) provide the recursions for the partial numerators and denominators of the socalled even parts of the continued fraction (4.15), for $\nu = 0$, and for the corresponding fractions for arbitrary ν.

It is, of course, the fact that the p's and the r's satisfy the same identities that permits the representation of any entry in the Padé table as a continued fraction. Many identities other than those already given can be formed by methods of elimination, and from these can be formed the continued fraction expansions following an arbitrary row or pair of adjacent rows, an arbitrary column or pair of adjacent columns, or an arbitrary diagonal (see [8] for a large collection of these and related identities).

5. Some Applications

One of the more elementary questions discussed in the literature, and generally laboriously, is that of the existence of a continued fraction expansion of specified form, say (3.9). The answer to this is given immediately by (3.8). Equally laborious is the usual text-book discussion of the representation of a power series by a continued fraction of specified form. Necessary and sufficient conditions for the possibility can be expressed in terms of the nonvanishing of certain Hankel determinants, assuming the existence of the relevant entries in the Padé table, and the terms of the fraction, when it exists, come from the coefficients in the recursions discussed above.

Possibly less obvious but no less real is the relevance of this development to localization problems for the zeros of a polynomial. Typically this means counting the zeros within a given circle in the complex plane, or, equivalently, in a given half plane. It is possible to split the polynomial into two parts, according to the location of the bounding line or circle, and the count can be made by examining the signs of certain bigradients formed out of their coefficients. The Routh-Hurwitz criterion is a well known special case. The proof usually proceeds from a definition of the Cauchy index for the quotient of the two polynomials, through the formation of a generalized Sturm sequence by means of the Euclidean algorithm, to a laborious analysis that relates the bigradients to the signs at $\pm \infty$ of the members of the Sturm sequence. But the results of this analysis are contained essentially in (3.8) and (3.11). Computationally it seems that Schur's use of the Euclidean algorithm is perhaps more efficient than the direct use of the bigradients, and this is the approach taken by Derwidué (for half

planes) and by Lehmer (for disks). But this does not negate the
theoretical importance of the bigradients themselves.

A rather more direct approach is given by Gantmacher in
his last chapter, which is further simplified by application of
Theorem 2. The step that leads up to this is in showing that if a
rational fraction is expanded in descending powers of z, and the
matrix of $H_0^{(\nu)}$ of sufficiently high order formed, then the signature of this matrix is equal to the Cauchy index of the fraction.

REFERENCES

1. Erwin H. Bareiss: Resultant procedure and the mechanization of the Graeffe process, JACM *7* (1960), pp. 346-86.

2. Maxime Bôcher: Introduction to higher algebra, New York, The MacMillan Co., 1924, pp. xi + 321.

3. Emile Borel: Leçons sur les fonctions méromorphes, Paris, Gauthier-Villars, 1903, pp. vii + 122.

4. G. Frobenius: Ueber Relationen zwischen den Näherungsbrüchen von Potenzreihen, J. Reine Angew. Math. *90* (1880), pp. 1 - 17.

5. F.R. Gantmaher: Teorija matric. Izdanie vtoroe. Dopolnennoe, Izdatel'stvo "Nauka", Moskva, 1966, pp. 576.

6. Thomas Muir: The theory of determinants in the historical order of development, London, Macmillan and Company, Ltd., v.1, 1906, pp. xi + 491; v. 2, 1911, pp. xvi + 475; v. 3, 1920, pp. xxvi + 503; v. 4, 1923, pp. xxvi + 508.

7. Nicola Trudi: Teoria dei determinanti e loro applicazioni, Napoli, Libreria Scientifica e Industriale de B. Pellerano, 1862, pp. xii + 268.

8. Peter Wynn: The rational approximation of functions which are formally defined by a power series expansion, Math. Comp. *14* 1960, pp. 147-86.

Addendum: After writing the above the authors discovered the paper by Netto referenced below in which two special cases, other than Borel's, of Theorem 2 are given. It is interesting to note that the proof used for the second extends quite easily to a proof of the general theorem as stated here. Netto further states Trudi's theorem (though without reference to Trudi) for the special case $m = n - 1$, and shows that the polynomials here designated φ_i and ψ_j are equal, respectively, to the polynomials $M_0^{(i)}$ and $L_0^{(j)}$, each multiplied by a suitable power of z and a certain constant, provided, of course, $S(z)$ is the expansion of g/f. Again this method of proof can be extended to show that each $M_\nu^{(p)}$ and $L_\nu^{(p)}$ can be expressed as bigradient polynomials similar in form to the φ_i and ψ_j.

Eugen Netto (1896) Zur Theorie der Resultanten. J. Reine Angew. Math. *116*: 33-49

Prof. A.S. Householder
Director, Mathematics Division

Dr. G.W. Stewart, III
Oak Ridge Gaseous Diffusion Plant

Oak Ridge National Laboratory
Oak Ridge, Tennessee 37830

M. A. Jenkins and J. F. Traub

An Algorithm for an Automatic General Polynomial Solver

Abstract

A general automatic equation solver should be based on a restriction-free mathematical algorithm. By this we mean the algorithm should be suitable for all polynomials and not depend on the properties of certain classes of polynomials. In this paper we will describe a restriction-free algorithm and discuss a program which implements it.

The algorithm consists of two stages. The First Stage may be viewed as a preprocessing step which guarantees that the Second Stage iteration will converge. The zeros are found one or two at a time and in increasing order of magnitude which guarantees stable deflation.

Let k be the number of distinct smallest zeros of equal magnitude. After a possible translation we may assume $k = 1$ for polynomials with complex coefficients or $k = 2$ for polynomials with real coefficients. We deal with polynomials with real coefficients, which is the harder case.

We summarize the major decisions which are all made automatically by the program:

a) When do we switch from Stage One to Stage Two?
b) Is $k = 1$, 2 or is $k > 2$?
c) If $k > 2$, by how much should we translate?
d) What value should be assigned to the initial approximation for the Stage Two iteration?
e) What is the termination criterion for the Stage Two iteration?

Decisions a, b, and d are made on the basis of the same calculation.

If a zero is "easy" to calculate, then Stage One is automatically abbreviated. Time is spent on the "hard" zeros. *One may view the technique as one which involves a spectrum of iteration methods with the appropriate one automatically selected.*

An ALGOL program implements the algorithm. Flow-charts for this program are given. The program has been tried on some of the hardest problems we could find including a 20-th degree polynomial of Wilkinson and a 36-th degree polynomial of Henrici. Numerical results are presented.

We consider the current computer program as a research program which demonstrates that it is feasible to make the major decisions automatically. Certain improvements will have to be made before the program could serve as a general library routine.

TABLE OF CONTENTS

		Page
1.	Introduction	153
2.	The Mathematical Algorithm	155
3.	Properties of the Mathematical Algorithm	157
4.	Decisions to be Made in the Program	161
5.	The Termination of Stage One	162
6.	The Translation of the Polynomial	165
7.	Scaling	166
8.	Termination of Second Stage Iteration	167
9.	Numerical Results	167
10.	Summary	173
11.	Acknowledgments	175
	Appendix - Flowcharts	176
	References	179

1. Introduction

A general automatic equation solver should be based on a restriction-free mathematical algorithm. By this we mean the algorithm should be suitable for all polynomials and not depend on the properties of certain classes of polynomials. In this paper we will describe a restriction-free algorithm and discuss a program which implements it.

The algorithm enjoys a basic simplicity and requires few decisions. We devise procedures by which the computer may automatically make the major decisions required. We do not concern ourselves here with programs used in an interactive environment. Routines to be used in such an environment might have different characteristics.

We summarize a few of the desirable characteristics of the algorithm. It is basically iterative with a preprocessing stage which guarantees that the iteration will converge. Often the most difficult problem associated with an iterative method is the value of the initial iterate. This is easy for us to handle because the mathematical algorithm will converge for essentially all initial approximations while our implementation of the algorithm actually supplies us with a good initial approximation. Multiple zeros require no special handling. Finally, the importance of finding the zeros in roughly increasing order of magnitude to ensure stable deflation has been stressed by Wilkinson [11, p. 465] who observes there seems no reliable method for ensuring this. Our algorithm does find the zeros in roughly increasing order of magnitude.

The last point merits some amplification. If the zeros are found in decreasing order of magnitude, then the backward deflation is stable. What is really crucial is that at each stage of the deflation either one of the smallest or one of the largest zeros is calculated.

The algorithm may be applied to polynomials with real or complex coefficients. Polynomials with complex coefficients are easier to deal with for the following reason. Let k be the number of distinct smallest zeros of equal magnitude. Although theoretically the kind of method which we will describe could be extended to handle zero distributions with any value of k, the simplicity of the implementation depends on k being small. In the case of a polynomial with complex coefficients we can, after a complex translation, ensure $k = 1$. For a polynomial with real coefficients, we are left with the cases $k = 1$ and $k = 2$ if we restrict ourselves to real translations.

The algorithm to be introduced in this paper is a member of a class of two-stage methods introduced by Traub. This type of method was first announced in [5]. The calculation of the largest zero of a polynomial was discussed in detail in [6] and global convergence was proven for a class of methods. For the largest zero the first stage involves the generation of G polynomials. The proof of global convergence of an algorithm for computing complex conjugate zeros was announced in [7] while the calculation of the smallest zero and of multiple zeros as well as the extension to analytic functions appears in [8].

The calculation of the smallest zero involves H polynomials. G polynomials and H polynomials have a simple relation and any result involving one can be translated into a result involving the other. Calculating the smallest zero first makes translation more effective. Hence, we shall be involved with H polynomials.

Bibliographic remarks and rather extensive bibliographies may be found in Traub [6], [8].

The papers cited above deal with finding one zero or a complex conjugate pair and focus on mathematical properties. In this paper, we focus on a particular algorithm out of a class of possible algorithms and discuss its feasibility as the basis for a general automatic equation solver.

2. The Mathematical Algorithm

Let

$$P(t) = \sum_{j=0}^{n} a_j t^{n-j}, \quad \begin{array}{l} a_0 = 1 \\ a_n \neq 0 \end{array}$$

be a polynomial with l distinct zeros ρ_i of multiplicity m_i.

Stage One

We generate a sequence of polynomials as follows. Let

$$H(0,t) = P'(t)$$

(2.1) $\quad H(\lambda+1,t) = \frac{1}{t}\left[H(\lambda,t) - \frac{H(\lambda,0)}{P(0)} P(t) \right], \quad \lambda = 0,1,\ldots,\Lambda.$

Observe that the polynomials are of degree at most $n - 1$.

Stage Two

Let k be the number of distinct zeros of smallest magnitude. If $k > 2$, translate the polynomial so that $k = 1$ or 2. Observe that the distinction between $k = 1$ and $k = 2$ is of importance only for the case of real coefficients.

We introduce the following notation to help us describe the Stage Two iteration. Let $h(t)$ be a polynomial of degree r. Then

$\bar{h}(t)$ is the polynomial $h(t)$ divided by the coefficient of t^r. Let t_0 be the initial iterate. Then we generate a sequence of iterates by $t_{i+1} = \psi_k(t_i, f)$ where f is the function whose zero we seek. (In this notation, Newton-Raphson iteration is defined by $\psi(t,f) = t - f/f'$.)

We can now give the formulas of the iteration functions for $k = 1$ and 2.

$k = 1$

Let
$$t_{i+1} = \psi_1(t_i, f)$$
where
$$\psi_1(t,f) = t - f/f' ,$$
and
$$f \equiv V(\Lambda, t) = P(t)/\bar{H}(\Lambda, t) .$$

$k = 2$

Let
$$t_{i+1} = \psi_2(t_i, f)$$
where
$$\psi_2(t,f) = t - \frac{2f}{f' \pm \left[(f')^2 - 4f\right]^{1/2}}$$
and
$$f \equiv W(\Lambda, t) = P(t)/\bar{I}(\Lambda, t) ,$$
$$I(\Lambda, t) = \delta(\Lambda-1)H(\Lambda, t) - \delta(\Lambda)H(\Lambda-1, t)$$

with $\delta(\Lambda)$ the coefficient of t^{n-1} in $H(\Lambda, t)$. Let the zero be labeled α. If $k = 1$, the polynomial $P(t)/(t-\alpha)$ is formed. If $k = 2$, the polynomial $P(t)/[(t-\alpha)(t-\bar{\alpha})]$ is formed. We then return to Stage One with the new polynomial.

3. Properties of the Mathematical Algorithm

We shall now state a number of results which exhibit the power of the mathematical algorithm. Results analogous to results stated for the case $k = 1$ may be found in Traub [6]. Proofs of results for the case $k = 2$ will appear elsewhere. The notation is the same as in Section 2. We shall use λ as a running index and Λ as a fixed integer.

Most of our results follow from the formula given in

Theorem 1. For all λ,

(3.1)
$$\frac{H(\lambda,t)}{P(t)} = \sum_{i=1}^{l} \frac{m_i \rho_i^{-\lambda}}{t-\rho_i}.$$

The key property of $H(\lambda,t)$ is given in the following two theorems.

Theorem 2. Let $|\rho_1| < |\rho_i|$, $i > 1$. Then for all finite t,

$$\lim_{\lambda \to \infty} V(\lambda,t) = t - \rho_1.$$

Theorem 3. Let $|\rho_1| < |\rho_i|$ and $|\rho_2| < |\rho_i|$, $i > 2$. Then for all finite t,

$$\lim_{\lambda \to \infty} W(\lambda,t) = (t-\rho_1)(t-\rho_2).$$

Note that the hypothesis of Theorem 3 includes the case $k = 2$. Generalizations of these theorems hold for the case of k smallest zeros in magnitude. The rate of convergence is of Bernoulli type.

Results concerning the zeros and poles of $V(\lambda,t)$ and $W(\lambda,t)$ are given in the following group of theorems. Note that no restrictions have been imposed on the multiplicities of the zeros of P. The following theorem is a generalization of the statement that the rational function P/P' has only simple zeros. Observe that $V(0,t)$ is proportional to P/P'.

<u>Theorem 4</u>. *For all finite* λ, $V(\lambda,t)$ *and* $W(\lambda,t)$ *have only simple zeros and these are the zeros of* P.

<u>Theorem 5</u>. *Let* $|\rho_1| < |\rho_i|$, $i > 1$. *Let* K_1 *be the union of circles with arbitrarily small fixed radii centered at the* ρ_i, $i > 1$. *Then for* λ *sufficiently large, the poles of* $V(\lambda,t)$ *are contained in* K_1.

<u>Theorem 6</u>. *Let* $|\rho_1| < |\rho_i|$, $|\rho_2| < |\rho_i|$, $i > 2$. *Let* K_2 *be the union of circles with arbitrarily small fixed radii centered at the* ρ_i, $i > 2$. *Then for* λ *sufficiently large, the poles of* $W(\lambda,t)$ *are contained in* K_2.

We now state some theorems concerning the iteration functions ψ_1 and ψ_2. As usual we define the order of the iteration as follows. Let $t_i \to \alpha$. Then if there exists a constant p and a nonzero constant C such that

$$\lim_{t_i \to \alpha} \frac{(t_{i+1} - \alpha)}{(t_i - \alpha)^p} = C$$

then p is called the order and C the asymptotic error constant. For our iteration functions, $C = C_k(\lambda)$. We then have

<u>Theorem 7.</u> ψ_1 and ψ_2 *are second order iteration functions. Furthermore,*

(3.2) $$\lim_{\lambda \to \infty} C_k(\lambda) = 0, \quad k = 1,2 \; .$$

We comment on this result. The iteration is done for a fixed value of $\lambda = \Lambda$. Theorem 7 shows that if Λ is large, $C_k(\Lambda)$ will be small. Hence, although the iteration is of second order, the error at each iteration will be the product of three small numbers and hence will appear faster than the usual quadratic convergence. Additional discussion of $C(\Lambda)$ may be found in Traub [6, Section 6] and [8, Section 7].

The speed of convergence is illustrated by the following simple example which we take from Traub [6]. In this example an earlier program is used which calculates the biggest zero first and which does not make decisions automatically.

Let
$$P(t) = t^4 - 46t^3 + 528t^2 - 1090t + 2175.$$

The largest zero is 29. Take $\Lambda = 16$. Let

$$t_0 = 100\ 000.$$

Then

$$t_1 = 28.99963$$

$$t_2 = 28.9999999999997$$

We hope the following discussion will offer some insight into the choice of ψ_1 and ψ_2 and will clarify the reason why (3.2) holds.

Let $v(t) = (t-\rho_1)$, $w(t) = (t-\rho_1)(t-\rho_2)$. Then Theorems 2 and 3 may be restated as

$$\lim_{\lambda \to \infty} V(\lambda, t) = v(t),$$

$$\lim_{\lambda \to \infty} W(\lambda, t) = w(t).$$

Now,

$$\rho_1 \equiv \psi_1[t, v(t)],$$

$$\rho_1, \rho_2 \equiv \psi_2[t, w(t)].$$

Thus if $k = 1$ and we have taken λ to ∞, then starting with any t_0, the iteration with ψ_1 would have delivered the exact zero in one iteration and analogously for $k = 2$ and ψ_2.

We now state theorems on global convergence of the iterations defined by ψ_1 and ψ_2.

<u>Theorem 8.</u> Let $|\rho_1| < |\rho_i|$, $i > 1$. Let t_0 be an arbitrary point in the extended complex plane such that $t_0 \neq \rho_i$, $i > 1$ and let $t_{i+1} = \psi_1(t_i, V)$. Then for Λ sufficiently large but fixed, the sequence t_i is defined for all i and $t_i \to \rho_1$.

<u>Theorem 9.</u> Let $|\rho_1| < |\rho_i|$, $|\rho_2| < |\rho_i|$, $i > 2$. Let t_0 be an arbitrary point in the extended complex plane such that $t_0 \neq \rho_i$, $i > 2$ and let $t_{i+1} = \psi_2(t_i, W)$. Then for Λ sufficiently large but

fixed, the sequence t_i is defined for all i and $t_i \to \rho_1$.

These theorems require a few words of comment. The formulas for ψ_1 and ψ_2 as given above make it appear as if these functions are not defined at ∞. However, ψ_1 and ψ_2 may be rewritten so that they are defined at ∞. (Observe that this is not a property shared by the Newton-Raphson iteration function.)

The iteration ψ_2 is multivalued because of the $\pm$ sign. However, a strategy is available for making the iteration converge to either ρ_1 or ρ_2. A discussion of this in a somewhat different setting is given by Traub [8, Section 12].

A proof of Theorem 8 for the case where P has only simple zeros is given by Traub [6, pp. 121-123]. The extension to multiple zeros is not difficult.

These theorems show that if we apply our two stage algorithm to any polynomial, with perhaps a translation to ensure $k = 1$ or 2, then provided Λ is sufficiently large, the mathematical algorithm is guaranteed to converge. For the remainder of this paper we discuss the implementation of this algorithm on a digital computer.

4. Decisions to be Made in the Program

We enumerate the major decisions that have to be made automatically by a program implementing this algorithm. A number of the decisions are not crucial and are made on an ad hoc basis. Other decisions are crucial and are made on the basis of certain calculations.

We summarize the major decisions to be made in the calculation of each zero or pair of zeros:

a) What is Λ, the value of λ for which we terminate Stage One and switch to Stage Two?
b) Is $k = 1, 2$, or is $k > 2$?
c) If $k > 2$, by how much should we translate?
d) What value should be assigned to t_0, the initial iterate for Stage Two?
e) What is the termination criterion for the Stage Two iteration?

Decisions a,b, and d are made as the result of the same calculation. Indeed, Stage One is terminated when k can be determined as equal to 1 or 2. If such a determination cannot be made by the time that λ has reached a certain value λ_f, a translation is carried out.

Decision d, which is often the most difficult decision to make, is available here as a byproduct. However, Theorems 8 and 9 show that the choice of t_0 is not crucial.

The methods for making decisions c and e are described in Sections 6 and 7, respectively.

5. The Termination of Stage One

If there are k smallest zeros in magnitude, then for λ sufficiently large

$$\frac{H(\lambda,t)}{P(t)} \sim \sum_{i=1}^{k} \frac{m_i \rho_i^{-\lambda}}{t-\rho_i}.$$

Hence $H(\lambda,t), \ldots, H(\lambda+k,t)$ will approximately satisfy the k-th order recurrence

(5.1) $$\sum_{i=0}^{k} c_{k-i} H(\lambda+i,t) = 0, \quad c_0 = 1,$$

where the c_i are related to the zeros of P by

$$\sum_{i=0}^{k} c_{k-i} t^i = \prod_{i=1}^{k} (t-\rho_i^{-1}) .$$

We wish to test the hypothesis that the $H(\lambda,t)$ satisfy a recurrence of the form (5.1) with $k = 1$ or 2. For a fixed value of λ we test the hypothesis $k = 1$ and if that seems to be false we test $k = 2$. If that is also false, we increase λ by a certain amount and test again. This is continued until a preset upper limit of λ is reached. At that point the polynomial is translated and we start again.

We describe the test for $k = 2$. The test for $k = 1$ is the appropriate simplification. Let

$$\delta(\lambda) = \sum_{i=1}^{l} m_i \rho_i^{-\lambda}$$

denote the leading coefficient of $H(\lambda,t)$. Let h_λ denote the vector of coefficients of the polynomial $H(\lambda,t)$. We apply two tests, the second being more expensive than the first and applied only if the first is passed.

We first test for a second order scalar recurrence. If this is passed, we test for a vector recurrence.

Let

$$D(\lambda) = \begin{vmatrix} \delta(\lambda+2) & \delta(\lambda+1) \\ \delta(\lambda+1) & \delta(\lambda) \end{vmatrix} ,$$

$$R(\lambda) = \frac{D(\lambda+1)}{D(\lambda)} .$$

If $\delta(\lambda)$ satisfies a second order scalar recurrence, $R(\lambda)$ converges. Hence the first test is

$$(5.2) \qquad \frac{|R(\lambda+1) - R(\lambda)|}{R(\lambda+1)} < \varepsilon.$$

If this test is passed, we test for the vector recursion as follows.

Choose $\hat{c}_1$, $\hat{c}_2$ so that the quantity

$$(5.3) \qquad r_\lambda = k_{\lambda+1} + \hat{c}_1 k_\lambda + \hat{c}_2 k_{\lambda-1}$$

is minimized in the L_2 norm and test if

$$(5.4) \qquad \frac{||r_\lambda||_2}{||k_\lambda||_2} < \varepsilon.$$

That is we solve a $n \times 2$ least squares problem. If (5.4) holds, we calculate the zeros q_1 and q_2 of $t^2 + \hat{c}_1 t + \hat{c}_2$ and use q_1^{-1} or q_2^{-1} as the value of the initial iterate t_0 in the second stage.

A similar least squares technique is used by Zurmühl [12] whose purpose it is to calculate approximations of equimodular eigenvalues using vectors generated by the power method. Zurmühl proves that he obtains Rayleigh approximations in this way. He does not use this as a criterion for termination.

We emphasize that the test tells us

a) When to switch from Stage One to Stage Two.

b) The value of k.

c) The value of t_0.

6. The Translation of the Polynomial

If the tests described in Section 5 have not been passed by a certain value of $\lambda = \lambda_f$, we translate the polynomial. Since we wish to calculate the zeros in the order of increasing magnitude, we do not want to shift by an amount which would place near the origin a zero with a significantly larger modulus than the smallest zeros. To ensure that this does not happen, we calculate a lower bound on the moduli of the zeros and use this quantity for a shift along the real axis. The lower bound we use, (Marden [4, p. 98]), is the unique positive zero of

$$(6.1) \qquad Q(t) = - \sum_{j=0}^{n-1} |a_j| \, t^{n-j} + |a_n|.$$

The positive zero of (6.1) is easily found by Newton-Raphson iteration. It need not be found very accurately.

One may construct examples which show that the smallest zero of the translated polynomial need not be the translated smallest zero of the original polynomial. However, these examples are based on near equimodular zero distributions and hence will not effect our statement that we can ensure stable deflation. However, this shows that the translated polynomial may have more than two equimodular smallest zeros. In the program we try shifts in both directions. An example of this may be seen in Example 3 of Section 8.

We now show how we may perform the translation and still use the original polynomial in the Stage Two iteration. This is clearly desirable numerically.

Let $p(t) = P(t-s)$ be the translated polynomial. With $s > 0$, this means the zeros of P are shifted s units to the right.

Let $\{h(\lambda,t)\}$ be the "H polynomial sequence" for $p(t)$ and let η_1 be the smallest zero of $p(t)$. [We assume for simplicity of exposition that $p(t)$ has a smallest zero. This is not essential.] Let $e(\lambda,t) = h(\lambda,t+s)$. Let

$$\sigma(\lambda,t) = \frac{P(t)}{\bar{e}(\lambda,t)}.$$

Then we have

<u>Theorem 10</u>. *Let* $p(t)$ *have zeros* η_i *with* $|\eta_1| < |\eta_i|$, $i > 1$. *Let* $t_{i+1} = \psi_1(t_i,\sigma)$. *Then* $t_i \to \eta_1 - s$.

An analogous result holds if the translated polynomial has two smallest zeros. After the zero has been calculated, the deflation is carried out in the original polynomial. Although this scheme requires two translations, it is not sensitive to roundoff error since the iteration is done in the original polynomial and hence the translations need not be done in higher precision than the rest of the calculation.

7. Scaling

We turn first to the scaling of the H polynomials. From (3.1), we see that as λ increases the coefficients of the $H(\lambda,t)$ grow or diminish depending on whether $|\rho_1| < 1$ or $|\rho_1| > 1$. Thus the coefficients must be periodically scaled. Scaling is done using a power of the radix (a power of 8 on the Burroughs B5500). An alternative scaling is discussed by Traub [6, Section 9].

We scale the least squares problem by replacing the problem of minimizing (5.3) with the problem of minimizing

(7.1) $$\underset{\sim}{g}_\lambda = D\underset{\sim}{k}_{\lambda+1} + \hat{c}_1 D\underset{\sim}{k}_\lambda + \hat{c}_2 D\underset{\sim}{k}_{\lambda-1}$$

where D is the diagonal matrix whose j-th diagonal element is the average of the j-th component of $\underset{\sim}{k}_{\lambda-1}$, $\underset{\sim}{k}_\lambda$, $\underset{\sim}{k}_{\lambda+1}$. If this scaling is not done, the minimization of (5.3) may reflect only one very large component which might satisfy a three-term recurrence whereas the vector does not.

8. Termination of Second Stage Iteration

A problem common to all iterative methods is when to terminate the process. Generally, the decision to terminate has been based on an ad hoc criterion such as $|t_{i+1}-t_i|/|t_i| < \varepsilon$, with the parameter ε chosen a priori.

In our program we terminate iteration on the basis of a technique due to Kahan. He derives an a posteriori bound on the roundoff error in evaluating a real polynomial at a real point and suggests that iteration be stopped when the computed value of the polynomial is less than a small integer multiple of this bound. Kahan's technique appears without explanation in Kahan and Farkas [3]. Adams [1] analyzes the case of a real polynomial evaluated at a complex point and shows that the bound is tight enough so that the iteration is not stopped prematurely. Our experience with this leads us to the conclusion that it is an excellent way in which to terminate the second stage.

9. Numerical Results

An ALGOL program has been written for the Burroughs B5500 to test the algorithm described in this paper. The Stage One calculation is done in single precision (13 octal digits) and the

Stage Two iteration is done in double precision (26 octal digits).

The program makes good use of the recursive facility of ALGOL. Flowcharts of the program may be found in Appendix A.

The procedures for automatically making the important decisions listed in Section 4 have been described earlier. A number of other parameters, whose values do not play a critical role, are chosen on an ad hoc basis. We discuss the values assigned these parameters in our program.

The switchover test is applied each time that λ has been increased by 4. The maximum value of λ permitted in Stage One is $\lambda = 200$. If the switchover test has not been passed by this time, we translate.

The maximum number of iterations permitted in Stage Two is 6. This choice is based on the assumption that there will be at least 1 correct figure in the initial approximation and with quadratic convergence 6 iterations will produce a double precision answer. (Double precision on the Burroughs B5500 is 23 decimal digits.)

The number ε appearing in (5.2) and (5.4) is initially set at $.001$. If the switchover test is passed and then iteration does not converge we replace ε by $\varepsilon/10$ and restart the Stage One calculation.

If the switchover test is not passed for $\lambda = 200$ and iterations following translation in both directions fail, then we increase the upper limit on λ and restart Stage One with ε replaced by 10ε.

We turn to three numerical examples. We tried the program on some of the hardest problems we could find. Examples of how this type of method does on simple problems may be found in Traub [6, Appendix A].

What are hard problems for zero finders? Multiple and near multiple zeros cause difficulty for most methods. Wilkinson [9] points out the difficulty in solving a polynomial whose zeros lie in an arithmetic progression. Equimodular and near equimodular zeros are difficult for methods involving zero separation such as Graeffe's method and power methods. Since Stage One of our algorithm can be interpreted as a power method, this presents us with our hardest problem.

E x a m p l e 1.

This is the 20-th degree polynomial with zeros at *1,2, ..., 20* discussed by Wilkinson. All the zeros are found to at least *10* decimal places of accuracy. Table I gives the zeros in the order in which they were found. Note that the zeros were calculated in strictly increasing order and that except for the zeros at *17* and *18* the value of Λ required to pass the switchover test increases as the ratio of the smallest to the next-to-smallest zero increases. As this ratio increases, the initial approximations become less accurate but in all cases are within 2 % of the zero. Thus our automatic switchover criterion is working very well. Since $k = 1$, ψ_1 is used throughout.

E x a m p l e 2.

This is the 19-th degree polynomial whose zeros are *.025 $\pm$ .035i, -.04 $\pm$ .03i, 27 $\pm$ .37i , -.4 $\pm$ .3i, 2.9 $\pm$ 3.9i, -4 $\pm$ 3i, 10 $\pm$ 2i, -20, 20, 30, 30, 30.* The first six pairs of complex conjugate zeros were chosen to test how the algorithm behaves with complex zeros of nearly equal modulus which are not clustered. The zeros at *20* and *-20* test the algorithm on equal

Table I

Zero	Λ	Estimate from Stage One	Number of Iterations
1	12	1.0002	2
2	16	2.0014	2
3	16	3.0096	3
4	20	4.011	3
5	20	5.026	3
6	24	6.025	3
7	24	7.042	3
8	24	8.064	3
9	28	9.055	3
10	28	10.075	3
11	28	11.097	3
12	28	12.12	3
13	28	13.15	4
14	32	14.12	4
15	32	15.15	4
16	32	16.17	4
17	28	17.25	4
18	4	18.14	5
19,20	Found Directly From the Final Quadratic Factor.		

real roots with opposite sign and the triple zero at 30 tests the behavior of multiple zeros. Table II gives the results in the order the zeros were found. The zeros are found by the algorithm to eleven significant figures except for the multiple zero at 30 for which the last two approximations agree only to 7 significant figures. This is all one expects for a triple zero. The results

Table II

Zero	Modulus	Λ	Estimate From Stage One	Number of Iterations
.025± .035i	.0430	68	.0250015 ± .0350012i	3
-.04 ± .03i	.0500	8	-.03999988± .0300002i	2
.27 ± .37i	.458	100	.26997 ± .36993i	3
-.4 ± .3i	.500	8	-.400002 ± .299996i	2
-4. ±3.i	5.00	12*	-4.0005 ±3.0003i	3
2.9 ±3.9i	4.86	12	2.9004 ±3.90009i	3
10. ±2.i	10.198	20	9.996 ±1.996i	3
-20.	20	12	-20.0089	2
20.	20	4	20.0000006	2
30.	30	4	29.999999995	0
30,30		Found Directly From the Quadratic		

*Switchover test not satisfied at $\lambda = 200$. Shift the zeros by and the test is passed with $\Lambda = 12$.

show that as the ratio of the smallest modulus to the next-to-smallest modulus increases, the switchover value Λ increases. For the first two pairs of zeros the ratio is *.860* and Λ is *68*. For the next two pairs the ratio is *.916* and Λ is *100*. For the last two pairs in this group of zeros the ratio is *.972* and the test is not passed for any $\lambda \leq 200$. In this case the program automatically shifts the zeros by *2.93*, the test is now passed with $\Lambda = 12$, and the algorithm converges to the zero at *-4 + 3i*.

Example 3.

This is the 36-th polynomial whose coefficients were chosen randomly by Henrici and Watkins [2]. All its zeros lie close to the unit circle which make this example a difficult one for our algorithm. The polynomial is $P = -9265.3t^{36} + 6468t^{35} - 42.015t^{34} + 70.311t^{33} + 3072.4t^{32} + 2.953t^{31} + 5.6163t^{30} + 870.73t^{29} - 7.9141t^{28} - 74.110t^{27} - 22.964t^{26} + 9.2252t^{25} - 2.4987t^{24} - 39.063t^{23} + 6.5810t^{22} - 6.8461t^{21} - 7.8867t^{20} - 32.151t^{19} - 34.637t^{18} + 67.916t^{17} - 390.57t^{16} + 60.247t^{15} + 265.74t^{14} - 453.86t^{13} - 7015.6t^{12} - 309.67t^{11} - 2.0574t^{10} - 85.581t^{9} - 99.394t^{8} - 20.775t^{7} + 49.225t^{6} + 3924.5t^{5} - .083830t^{4} + 73.941t^{3} + 0.049060t^{2} + 88.312t - 993.56.$

Many of the zeros required shifts before they were found. The column "number of shifts" in Table III has the following meaning. If the entry is *1* a shift of the zeros to the right has succeeded. If *2*, then the first shift has failed to produce a polynomial whose smallest root or pair of roots could be found, but a subsequent shift of the zeros to the left succeeded. If the entry is 2^+, then both the shifts have failed and the original Stage One calculation has been restarted. Note that one of the largest zeros has been calculated first. Since all the zeros are of comparable magnitude this does not cause trouble during deflation. Purification in the original polynomial produces no change in the approximate zeros in the 10 significant figures which are quoted in Table III.

Table III

Zero	Modulus	Number of Shifts	Λ	Number of Iterations
-.9828508293 ± .1124801357i	.989	1	44	3
-.5871954694 ± .4819686407i	.760	1	48	3
.2208328178 ± .7000044671i	.734	0	60	3
.8127598823 ± .0614797464i	.815	0	96	3
-.8845981101 ± .3015196625i	.935	1	108	4
-.8372852532 ± .3962960916i	.926	1	72	3
-.7466486856 ± .6041234511i	.960	1	144	4
.6330847696 ± .6931412366i	.939	2	92	4
-.2094825011 ± .8999616203i	.924	0	152	3
-.5831113093 ± .7852246824i	.978	1	184	4
-.3772294487 ± .8980728714i	.974	1	96	4
-.0725823577 ± .9908757215i	.994	1	164	4
.5788786350 ± .7720099987i	.965	0	120	3
.3926236740 ± .9249986413i	1.005	2+	384	5
.1585296127 ± .1000329321i	1.013	1	68	4
.8264593743 ± .5773500167i	1.008	0	188	5
.9538469054 ± .3607127224i	1.020	0	92	4
1.053012577 ± .0877037032i	1.057	Found Directly From the Quadratic		

10. Summary

Our major conclusion is that the algorithm described here can be used as the basis for a program which automatically calculates all the zeros of a polynomial and finds them in roughly increasing order of magnitude.

It is clear that our program could not compete in terms of computer time with a program which simply always uses Muller or Newton-Raphson iteration, when these iterations converge. Note, however, that if a problem is easy, then the switchover from Stage One to Stage Two is made early. Hence, easy problems are handled relatively cheaply. Time is spent on the hard zeros. *One may view the technique as one which involves a spectrum of iteration functions with the appropriate iteration automatically selected.*

Our program is designed to be a general polynomial solver. It must be able to handle all polynomials. If one knows a priori that one is dealing with a special polynomial, such as one with all distinct real zeros, then Newton-Raphson or Laguerre iteration may be used safely. Furthermore, a computer library should probably contain special routines for handling quadratic, cubic, and perhaps quartic equations. A general polynomial solver is needed to handle the cases where special properties of the polynomial are not known or if there is to be only one polynomial solver in the library.

The only difficult case for our algorithm is when there are near equimodular zeros such that our translations don't break the near equimodularity. We are studying methods for handling this problem either by a suitable modification of the algorithm or by switching to another method in case this difficulty occurs.

We are considering the feasibility of using complex translations even in the case of real polynomials. This would mean only the case $k = 1$ need be considered. It would also offer greater flexibility in translation.

Our program is incomplete in that no attempt is made to give a posteriori estimates of how good the calculated zeros are.

The numerical examples exhibited here, as well as other examples we have run, indicate that the switchover test works quite efficiently. In almost every case the test is not passed until λ is large enough so that the Stage Two iteration converges. The value of

λ at the switchover point increases as the ratio of the magnitude of the smallest zero to the magnitude of the next smallest zero increases. Usually the approximation from Stage One to Stage Two is good to between two to four figures which indicates the first stage has not been carried too far.

Observe that very few iterations are required in Stage Two as we would expect from the discussion of Section 3. Our numerical results confirm Wilkinson's conclusion [10, pp. 65] that there is little to be gained by purification in the original polynomial *provided that the zeros have been deflated in the proper order*. We have not found a single case where a zero is significantly improved by purification. This also indicates that our procedure for terminating Stage Two is working well.

11. Acknowledgments

It is a pleasure to express our appreciation to Professor W. Kahan of the University of Toronto who made many valuable comments and suggestions on the work reported in this paper. One of his contributions was to suggest the method of "double translation". Much of the research for this paper was done while one of us (JFT) was enjoying the hospitality of Professor G.E. Forsythe's Computer Science Department at Stanford University.

Appendix A

Flowcharts

These flowcharts are intended only to give the general flow of the program.

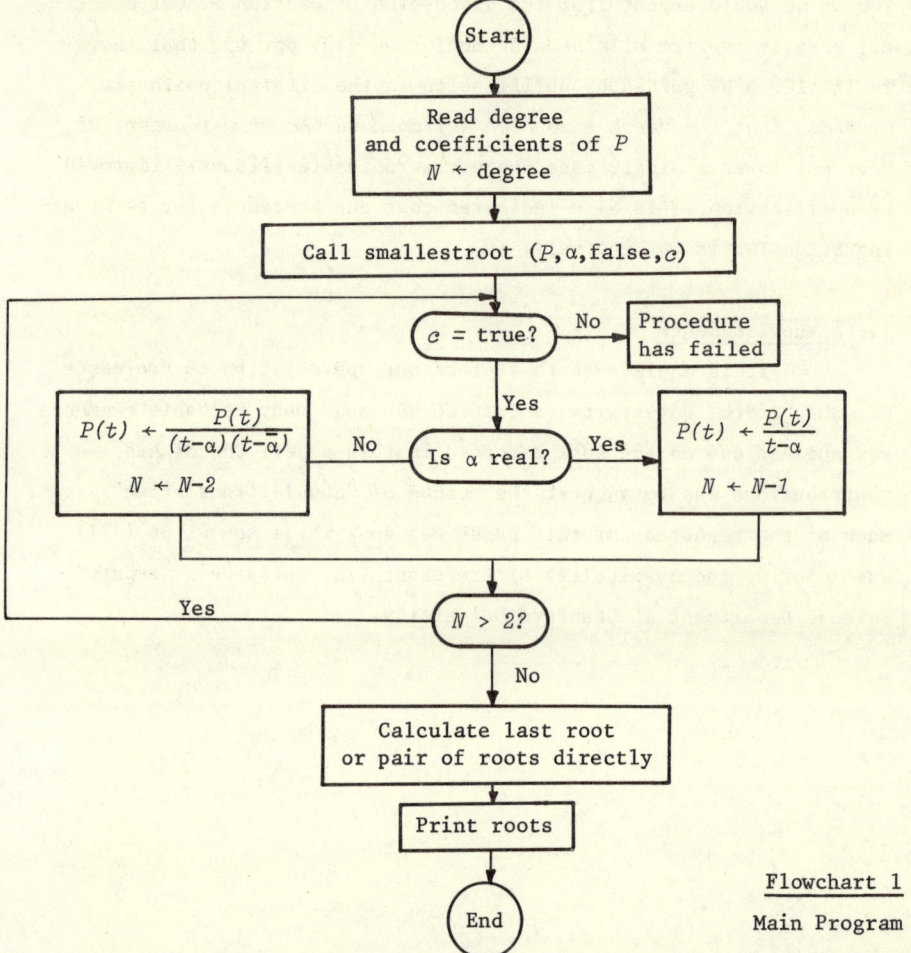

Flowchart 1
Main Program

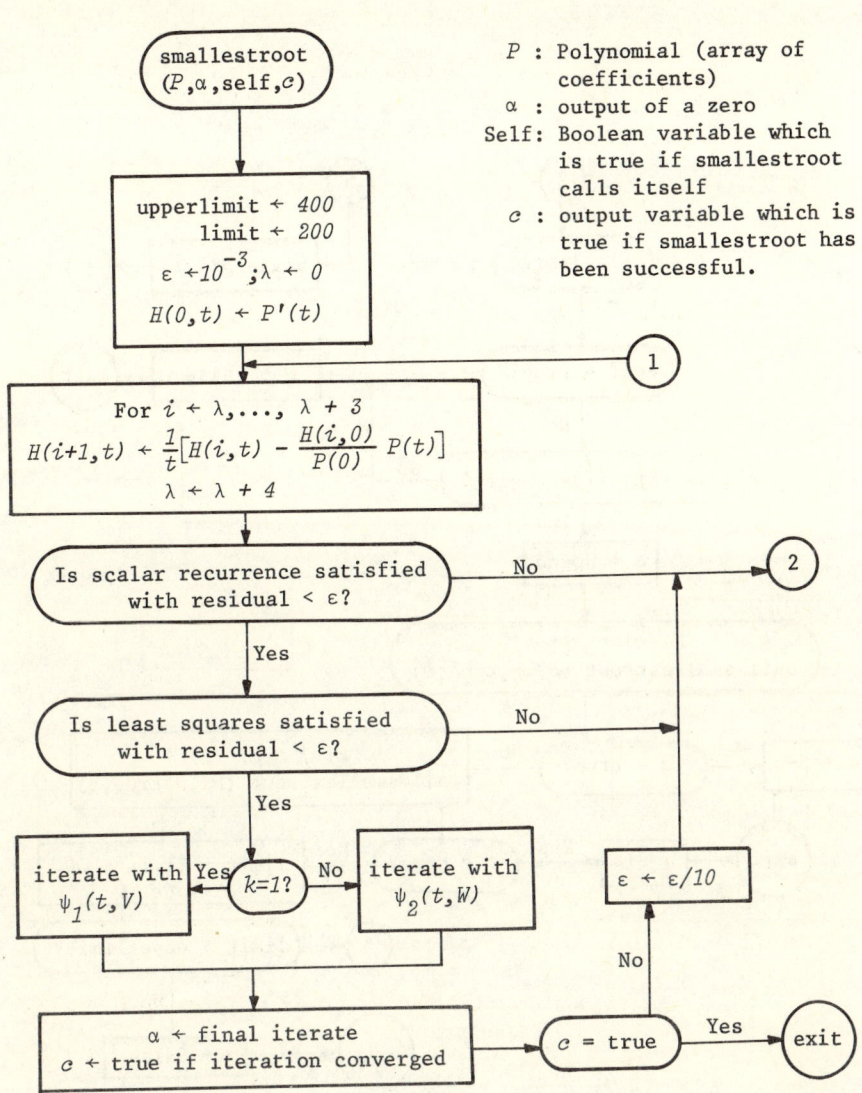

Flowchart 2

Procedure smallestroot

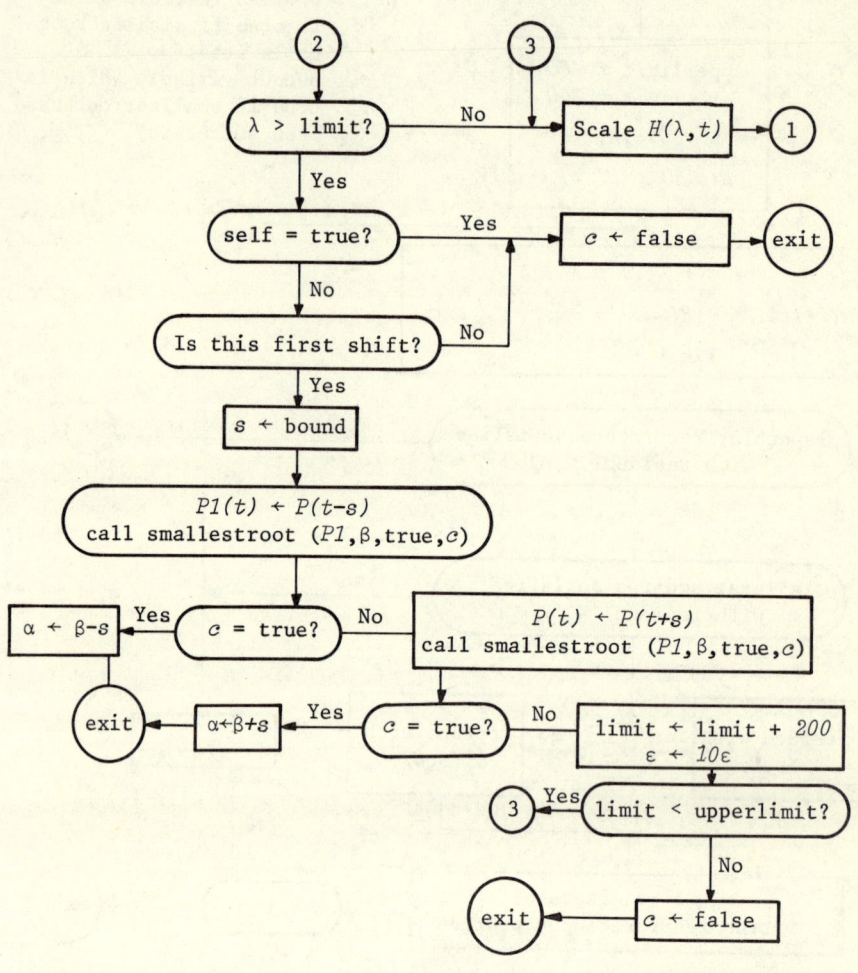

Flowchart 3
Procedure smallestroot (continued)

REFERENCES

1. D. Adams: A Stopping Criterion for Polynomial Root Finding. Comm. ACM *10* (1967), pp. 655-658. Also available as Technical Report 55, Computer Science Department, Stanford University.

2. P. Henrici and Bruce O. Watkins: Finding Zeros of a Polynomial by the Q-D Algorithm. Comm. ACM *8* (1965), pp. 572-573. See also Richard F. Thomas,Jr., Corrections to Numerical Data on Q-D Algorithm Comm. ACM *9* (1966), p. 322.

3. W. Kahan and I. Farkas: Algorithm 168 and Algorithm 169. Comm. ACM *6* (1963), p. 165.

4. M. Marden: The Geometry of the Zeros of a Polynomial in a Complex Variable. Amer. Math. Soc., Providence, Rhode Island, 1949.

5. J.F. Traub: A Class of Globally Convergent Iteration Functions for the Solution of Polynomial Equations. Proc. IFIP Congress 65, Vol. 2, pp. 483-484. Spartan Books, Washington, D.C.

6. J.F. Traub: A Class of Globally Convergent Iteration Functions for the Solution of Polynomial Equations. Math. Comp. *20* (1966), pp. 113-138.

7. J.F. Traub: Proof of Global Convergence of an Iterative Method for Calculating Complex Zeros of a Polynomial. Notices Amer. Math. Soc. *13* (1966), p. 117.

8. J.F. Traub: The calculation of Zeros of Polynomials and Analytic Functions. Proceedings of Symposia in Applied Mathematics, Volume 19, Mathematical Aspects of Computer Science, pp. 138-152, Amer.Math.Soc., Providence, Rhode Island, 1967. Also available as Technical Report 36, Computer Science Department, Stanford University.

9. J.H. Wilkinson: The Evaluation of the Zeros of Ill-Conditioned Polynomials: Part I. Num. Math. *1* (1959), pp. 150-166.

10. J.H. Wilkinson: Rounding Errors in Algebraic Processes. Prentice-Hall, 1963.

11. J.H. Wilkinson: The Algebraic Eigenvalue Problem. Clarendon Press, 1965.

12. R. Zurmühl: Rayleigh-Näherungen für Simultan-Iteration an betragsgleichen Eigenwerten einer Matrix. ZAMM *42* (1962), pp. 210-213.

M.A. Jenkins
Stanford University
Stanford, Cal. 94305

Dr. J.F. Traub
Bell Telephone Laboratories
Murray Hill, N.J. 07971

I. Kupka

Die numerische Bestimmung mehrfacher und nahe benachbarter Polynomnullstellen nach einem verbesserten Bernoulli-Verfahren

1. <u>Einleitung</u>

Das Bernoulli-Verfahren zur Bestimmung von Polynomnullstellen, auf Daniel Bernoulli [1], 1732, zurückgehend, gehört zu den numerischen Methoden, deren Bedeutung mit dem Einsatz von Rechenautomaten gewachsen ist. Friedrich L. Bauer [3] hat hierauf 1954 hingewiesen und ein darauf aufbauendes algorithmisches Verfahren entwickelt, welches jedoch bei Vorkommen mehrfacher und nahe benachbarter Nullstellen versagt. Der Einbeziehung dieser numerisch ungünstigen Fälle sowie der Abgrenzung der dabei auftretenden systematischen Schwierigkeiten und deren Beseitigung durch den gezielten Einsatz von Rechenoperationen in mehrfacher Genauigkeit soll die vorliegende Untersuchung dienen.

2. <u>Grundoperationen des Verfahrens</u>

Gegeben sei das reelle Polynom

$$f(x) = p_0 x^n + p_1 x^{n-1} + \ldots + p_n, \quad p_0 = 1, \quad p_n \neq 0,$$

$p_1, p_2, \ldots, p_n$ endliche Dualbrüche mit maximal a Mantissenstellen in vollnormierter Darstellung und mit Exponenten zur Basis 2 aus einem gleichfalls beschränkten Exponentenbereich.

Die numerische Bestimmung mehrfacher und nahe benachbarter Polynomnullstellen nach einem verbesserten Bernoulli-Verfahren

Der Fall $p_0 \neq 1$ kann allgemein nicht hierauf zurückgeführt werden, lässt sich aber in einem abgewandelten Verfahren ebenfalls behandeln.

Gefragt ist nach den Nullstellen von $f(x)$ in einer Genauigkeit von $a-1$ Binärstellen der Mantisse.

Das Prinzip des Bernoulli-Verfahrens ist die Bestimmung der Nullstellen des Polynoms als Eigenwerte der zugehörigen Begleitmatrix

$$L = \begin{pmatrix} \cdot & 1 & \cdot & \cdot & \cdot \\ \cdot & \cdot & \cdot & & \cdot \\ \cdot & & & 1 & \cdot \\ \cdot & & & & 1 \\ -p_n & \cdot & \cdot & -p_2 & -p_1 \end{pmatrix}$$

nach dem Verfahren der iterierten Vektoren. Bei der Iteration mit Spaltenvektoren

$$u_i = \begin{pmatrix} q_{i-n+1} \\ \cdot \\ \cdot \\ q_{i-1} \\ q_i \end{pmatrix}, \quad u_0 := \begin{pmatrix} 0 \\ \cdot \\ \cdot \\ 0 \\ 1 \end{pmatrix}, \quad i = 0, 1, 2, \ldots,$$

wird bei jedem Schritt $u_{i+1} = L u_i$ nur die Komponente

$$q_{i+1} = p u_i, \quad p := (-p_n, \ldots, -p_2, -p_1),$$

berechnet. Von einem Vektor u_i ausgehend lässt sich nach $n-1$ solchen Schritten bereits der Vektor u_{2i+n-1} berechnen. Das beruht darauf, dass in der Gleichung

$$q_{i+j+1} = pL^i u_j$$

die Matrix L^i ersetzt werden kann durch die Matrix

$$Q_i = \begin{pmatrix} q_i & q_{i-1} & \cdot & q_{i-n+1} \\ 0 & \cdot & \cdot & \cdot \\ \cdot & \cdot & \cdot & \cdot & \cdot \\ \cdot & \cdot & \cdot & \cdot & q_{i-1} \\ 0 & \cdot & \cdot & 0 & q_i \end{pmatrix}.$$

Man kommt damit zu einer abgekürzten Bernoulli-Iteration

$$u_{2i+n-1} = Bu_i \ .$$

Jede Anwendung von B erfordert $2n^2-n$ Multiplikationen. In anderer Form hat F.L. Bauer in der erwähnten Arbeit die abgekürzte Iteration für Zeilenvektoren angegeben. Es gilt

$$B(cu_i) = c^2 Bu_i \ .$$

Dadurch wird die Einhaltung einer festen Grössenordnung während der Rechnung ermöglicht.

Sind x_j, $j = 1,2,\ldots,m$, die Nullstellen von $f(x)$, jeweils mit der Vielfachheit n_j, also $\sum_{j=1}^{m} n_j = n$, so gibt es die eindeutige Zerlegung

$$u_0 = \sum_{j=1}^{m} u_0^j$$

des Anfangsvektors u_0 in Hauptvektoren jeweils von der Stufe n_j, für die also gilt (E ist hier die n-dimensionale Einheitsmatrix)

$$(L - x_j E)^{n_j - 1} u_0^j \neq 0, \quad (L - x_j E)^{n_j} u_0^j = 0, \; j = 1, 2, \ldots, m.$$

Für die Vektoren

$$u_i^j := L^i u_0^j$$

gilt dieselbe Aussage. Offenbar ist

$$u_i = \sum_{j=1}^{m} u_i^j .$$

Die abgekürzte Bernoulli-Iteration B ist auch auf die Hauptvektoren u_i^j anwendbar. Allgemein gilt

$$B \sum_j c_j u_i^j = \sum_j c_j^2 u_{2i+n-1}^j .$$

Hierauf beruhen die von F.L. Bauer entdeckten Eigenschaften der Stabilität und Selbstkorrektur der abgekürzten Iteration. Die Iteration $u_{i+1}^j = L u_i^j$ ist im allgemeinen nicht stabil. Für $i \to \infty$ gilt im Falle

$$|x_1| = |x_2| = \ldots = |x_h| > |x_{h+1}| \geq \ldots \geq |x_m|$$

die asymptotische Gleichung

$$u_i \sim \sum_{j=1}^{h} u_i^j .$$

Vielfachheiten bei den dominanten Nullstellen $x_1, x_2, \ldots, x_h$ führen noch zu einer Verschärfung dieser Aussage. Zu deren Verständnis muss eine weitere Zerlegung von u_i^j vorgenommen werden. Es gilt

$$u_i^j = \sum_{s=1}^{n_j} e_{j,s} u_i^{j,s} ;$$

dabei sind $u_i^{j,s}$ iterierte Hauptvektoren der Stufe s zu x_j und besitzen die spezielle Gestalt

$$u_i^{j,s} = \begin{pmatrix} \binom{i}{s-1} x_j^i \\ \binom{i+1}{s-1} x_j^{i+1} \\ \cdot \\ \cdot \\ \binom{i+n-1}{s-1} x_j^{i+n-1} \end{pmatrix}, \; j=1,2,\ldots,m, \; s=1,2,\ldots,n_j ,$$

$i = 0, 1, 2, \ldots$.

Genügen die Nullstellen der Beziehung

$$|x_1| = |x_2| = \ldots = |x_{h'}| = \ldots = |x_h| > |x_{h+1}| \geq \ldots \geq |x_m|$$

nebst

$$n_1 = n_2 = \ldots = n_{h'} > n_{h'+1} \geq \ldots \geq n_h ,$$

so gilt für $i \to \infty$ die asymptotische Beziehung

$$u_i \overset{\sim}{-} \sum_{j=1}^{h'} e_{j,n_j} u_i^{j,n_j} .$$

Schliesslich ist für $i \to \infty$ jeder Hauptvektor u_i^{j,n_j} asymptotisch proportional zu dem Eigenvektor $u_0^{j,1}$, also gilt

$$u_i \overset{\sim}{-} \sum_{j=1}^{h'} f_{j,i} u_0^{j,1} .$$

Die Zahl h' kann mit Hilfe eines von A.C. Aitken [2] angegebenen Algorithmus' bestimmt werden. Diese Bestimmung werde für $f'(x)$, $f''(x)$ usw. solange wiederholt, bis sich für h' ein grösserer Wert ergibt als am Anfang. Die Ableitungen können hierin divisionsfrei verwendet werden. Die Häufigkeit, mit der dieselbe Zahl h' ermittelt wird, ist gleich n_j, $j=1,2,\ldots,h'$. Sind h' und $n_1 = n_2 = \ldots = n_{h'}$ bekannt, so kann aus der Linearkombination $\sum_{j=1}^{h'} f_{j,i} u_0^{j,1}$ ein Biorthogonalsystem gebildet werden, mit dessen Hilfe

$$u_0^{(1,\ldots,h')} := \sum_{j=1}^{h'} u_0^j$$

berechnet werden kann. Zur weiteren Aufspaltung dieses Anfangsvektors ist eine Translation $\bar{x} = x - d$ der Polynomvariablen erforderlich verbunden mit einer zugehörigen umkehrbaren Transformation

$$(\bar{u}_0^{(1,\ldots,h')}, \bar{L}) = T_d (u_0^{(1,\ldots,h')}, L) .$$

Für T_d kann ein einfacher divisionsfreier Algorithmus angegeben werden. Dieser muss mit einer Rechengenauigkeit von $n(a+1)$ Binärstellen durchgeführt werden, damit das Polynom nicht verfälscht wird. Bei der Translation mit ausschliesslich reellem d erhält man durch Aufspaltung die Anteile u_0^j zu allen reellen Nullstellen x_j und $u_0^{(j,k)}$ zu allen konjugiert komplexen Wurzeln x_j, $x_k = \bar{x}_j$ (der Querstrich bezeichnet hier das Konjugiertkomplexe). Die Nullstellennäherungen können schliesslich als Nullstellen der Näherungspolynome

$$\begin{vmatrix} x & 1 \\ q_i^j & q_{i-1}^j \end{vmatrix} \quad \text{bzw.} \quad \begin{vmatrix} x^2 & x & 1 \\ q_i^{(j,k)} & q_{i-1}^{(j,k)} & q_{i-2}^{(j,k)} \\ q_{i+1}^{(j,k)} & q_i^{(j,k)} & q_{i-1}^{(j,k)} \end{vmatrix}$$

bestimmt werden.

Die angegebenen Schritte lassen sich theoretisch zu einem Algorithmus zur Nullstellenbestimmung zusammenfassen. Bei der Anwendung spielt jedoch die Rechengenauigkeit eine wesentliche Rolle. Die folgenden Untersuchungen hierüber sind keineswegs vollständig, liefern aber schon eine Reihe verwertbarer Erkenntnisse.

3. <u>Numerische Dominanz und Auslöschung</u>

Unter der binären Grössenordnung $b(x)$, x reell, $x > 0$, werde diejenige ganze Zahl b verstanden, für welche

$$2^{b-1} \leq x < 2^b$$

gilt. Ist entweder $y = 0$ oder $y \neq 0$ und

$$b(|x|) - b(|y|) > a$$

so heisse, bei fest vorgegebenem a, x numerisch dominant über y, in Zeichen $x \succ y$.

Für den Bernoullischen Iterationsprozess B ist das Auftreten numerischer Dominanzen charakteristisch, gleichzeitig tritt aber als erschwerende Begleiterscheinung die Auslöschung führender Stellen auf. Beide Effekte sollen in Abhängigkeit vom Iterationsindex untersucht werden.

Die wichtigsten Dominanzen sind

I) die Dominanz aufgrund grösseren Betrages

$$x_j^{i+n-1} \succ x_k^{i+n-1} \quad \text{für} \quad |x_j| > |x_k|,$$

II) die Dominanz aufgrund höherer Hauptvektorstufe

$$\binom{i+n-1}{s-1} \succ \binom{i+n-1}{t-1} \quad \text{für} \quad s > t$$

und III) die Dominanz der Eigenvektorkomponente der Hauptvektoren höchster Stufe

$$1 \succ \frac{\binom{i+n-1}{n_j-1}}{\binom{i+n-2}{n_j-1}} - 1 \,.$$

Ist $1 - \frac{|x_k|}{|x_j|} \geq 2^{-a+1}$, so wird I) erreicht für $i+n-1 \geq 2^{2a-2}$.
Dafür sind $r \geq 2a-2$ Schritte der abgekürzten Bernoulli-Iteration ausreichend. (Allgemein ist $B^r u_0 = u_{i_r}$ mit $i_r = (2^r-1)(n-1)$.)

II) wird erreicht für $r \geq a+1$ und III) für $r \geq a+3$.

Für die bei der Berechnung der neuen Vektorkomponenten gemäss

$$u_{2i+n-1}^{j,n_j} = Bu_i^{j,n_j}$$

auftretende maximale Anzahl $g_j(i)$ ausgelöschter führender Binärstellen erhält man durch Abschätzen der binären Grössenordnungen für die dort auftretenden Grössen eine obere Grenze. Wegen der grösseren Einfachheit der Formeln sei x_j reell und $|x_j| \geq 1$ angenommen. Dann gilt

$$g_j(i) \geq g_p + g_j + g_j'(i) .$$

Hierin treten ein nur vom Polynom abhängiger Term

$$g_p = 2b(n) + b(\max_k |p_k|)$$

auf, ein von j aber nicht von i abhängiger Term

$$g_j = b(|x_j|^{n-n_j} |e_{j,n_j}|),$$

der übrigens gegenüber Transformationen $\bar{x} = cx$, $c \neq 0$, invariant ist, und der Hauptterm

190 Die numerische Bestimmung mehrfacher und nahe benachbarter
Polynomnullstellen nach einem verbesserten Bernoulli-Verfahren

$$g'_j(i) = b(\binom{i}{n_j-1}^2 : \binom{2i}{n_j-1}).$$

Es gilt

$$|x_j|^{n-n_j}|e_{j,n_j}| \leq \left|\frac{x_j}{d_j}\right|^{n-n_j}, \quad \text{falls} \quad |x_j-x_k| \geq d_j \quad \text{für } k \neq j.$$

Ist $\left|\frac{x_j}{d_j}\right| \leq 2^{a-1}$, so folgt

$$g_j \leq a(n - n_j).$$

Der Fall $|x_j/d_j| > 2^{a-1}$ bleibt hier ausser Betracht. Für den Hauptterm gilt unter der Nebenbedingung $i > 8n$

$$g'_j(i) \leq 2 - b((n_j-2)!) + (n_j-1)b(i).$$

Ist $n_j > 1$ und $i = i_r$ mit $r \leq a$, so gilt die anschaulichere Abschätzung

$$g_j + g'_j(i) \leq 1 + (n-1)a.$$

Zur Vermeidung der störenden Einflüsse durch die Stellenauslöschung ist die Rechengenauigkeit bei der Iteration B sukzessive zu erhöhen, so dass maximal halb so viele Stellen ausgelöscht werden wie in die Rechnung gerade einbezogen sind. Von Vorteil wäre eine numerische Durchführung des Verfahrens, bei der die Rechengenauigkeit dynamisch durch die auftretende Stellenauslöschung gesteuert wird. Die Voraussetzungen hierzu dürften erst bei den schnelleren und über mehr Speicher verfügenden Rechnern bestehen, die sich zur Zeit noch in oder am Ende ihrer Entwicklung befinden.

Ein Teil der hier beschriebenen Operationen, nämlich die Bernoulli-Iteration B, die Bestimmung der Anzahl h' der Nullstellen maximaler Vielfachheit unter den betragsdominanten und die Bestimmung der Vielfachheiten wurden auf der TELEFUNKEN-Digital-Rechenanlage T R 4 im Rechenzentrum der Universität Hamburg erprobt. Mit Polynomen bis zum Grade $n = 12$ und jeweils n-facher Genauigkeit ($a=35$) wurde ein "relatives Auflösungsvermögen" des Verfahrens von 2^{-34} erreicht.

LITERATUR

1. D. Bernoulli: Observationes de seriebus recurrentibus, Comm. acad.sc.Imp. Petropol., Tom. III, (1732) p. 85-100.

2. A.C. Aitken: On Bernoulli's numerical solution of algebraic equations, Proc. Roy. Soc. Edinb., Vol. 46, (1927) p. 289-305.

3. F.L. Bauer: Quadratisch konvergente Durchführung der Bernoulli-Jacobischen Methode zur Nullstellenbestimmung von Polynomen, S.B. Bayer.Akad.Wiss., Math.-N. Kl., (1954) S. 275-303.

I. Kupka
Rechenzentrum der Universität
Rothenbaumchaussee 81
D-2 Hamburg 13

D. H. Lehmer

Search Procedures for Polynomial Equation Solving

The problem of solving polynomial equations is not only quite old but is also slow to modify itself to the age of computing. The more popular methods in use by modern computers are simply adaptations of the classical hand methods of the last two centuries. Once we allow ourselves the generality of equations with complex coefficients there is not any set of n complex numbers that cannot be the "solution set" of some polynomial. Hence there is no reason why equation solving cannot be regarded as a search problem in two dimensions. In this light it is obvious that for a digital computer with a finite precision and a limited amount of time, the problem of locating n arbitrarily situated points is indeed difficult. We shall have to be content with approximate answers in the form of a number of regions of small diameter which cover the n points in the complex plane.

There are however a few useful techniques whose elaboration and realization can lead to workable algorithms worthy of the machine's virtuosity, and these we propose to discuss.

Notation. Let $z = x + iy$ be a complex variable and let $\|z\|$ be a non-negative norm having the usual properties

$$||z|| = 0 \leftrightarrow z = 0$$

$$||z_1 + z_2|| \leq ||z_1|| + ||z_2||$$

$$||cz|| = |c|\ ||z|| \qquad (c\ real)$$

$$||z_1 z_2|| \leq ||z_1||\ ||z_2||\ .$$

Let $f(z)$ be a non-constant function, analytic at all interior points of a region R in the z-plane.

In seeking zeros of $f(z)$ we are searching the region R of the complex plane for points z_j for which $||f(z_j)|| = 0$.

Above the region R we can imagine a surface

$$S = S(f)$$

whose height above the point z in R is precisely $||f(z)||$. The desired zeros of f are just those points where the surface $S(f)$ dips down to touch the z plane. These minimum points of S are the *only* minima that S has. Indeed, suppose that z' is any interior point of R for which $||f(z')|| \neq 0$, then for every z in a small circular neighborhood of z' we have the convergent expansion

(1) $$f(z) = f(z') + \frac{(z-z')^r}{r!} f^{(r)}(z') + \sum_{n=r+1}^{\infty} a_n (z-z')^n\ ,$$

where $f^{(r)}(z') \neq 0$, $r > 0$ since f is not a constant. If we set

$$k = \frac{f^{(r)}(z')}{r!f(z')}\ , \qquad \zeta = z - z'$$

the expansion (1) can be written

(2) $$f(z) = f(z')(1 + k\zeta^r) + O(\zeta^{r+1})$$

as $\zeta \to 0$. Let us choose

$$\arg \zeta = \frac{\pi - \arg k}{r} .$$

Then for ζ sufficiently small in absolute value

$$0 < 1 + k\zeta^r < 1.$$

Taking the norm of both sides of (2) and using the above properties of the norm we find

$$||f(z)|| < ||f(z')||$$

holds for some z in every sufficiently small neighborhood of z'. Hence $||f(z')||$ is not a minimum point of S.

One way then to find the zeros of f is to throw a handful of small spheres onto the surface S and wait for them to roll into the desired pits. As the saying goes, "let gravity do your work."

Theoretically it is really not this easy because there must be a little ball in the watershed of each pit. Also some of the little balls may come to rest on peaks or passes, i.e. points of unstable equilibrium, rather than pits. With digital computing, the probability of this is not exactly zero, as we shall see.

A method based on this idea and suggested by Ward [2] runs into just such an accident.

In describing this method we may employ the following analogy. A parachutist is dropped onto the surface $S(f)$ with instructions to proceed down hill from his landing point until he finds himself at the bottom of a pit. Being digital, rather than analog, the parachutist does not roll like a ball under the force of gravity, but instead takes steps of varying length h. More precisely, after arriving at any point P on S (either on foot or by air) his instructions call for an exploration of the elevation of each of the four points one step of length h away in the four cardinal directions from P. He then is to choose the lowest of these four points as his next position (he can take the north point in case of a tie), provided this point be below P; otherwise he is to shorten his step ($h/2$ replaces h) and proceed to survey his surroundings as previously. Hopefully he reaches a pit z, his step length having become sensibly zero. This pit is now removed by dividing out:

$$f(z) = (z - z_1)f_1(x)$$

and replacing $S(f)$ by $S(f_1)$. A new parachute drop is made as before, the process continuing until a specified number of pits have been found. In actual practice Ward takes for R the square of side 2 centered at $z = 0$ with sides parallel to the real and imaginary axes. Whenever the parachutist leaves the projection of R onto the surface $S(f)$ the search is abandoned and $f(z)$ is replaced by $f(2z)$ whose roots are half those of $f(z)$. The initial drop is at $z = 0$ and the initial step length is taken as $h = 1/4$. The norm used is

$$||z|| = |Re(z)| + |Im(z)| .$$

As Ward points out, the method fails for

$$f(z) = 1 + z^4$$

since the initial point $z = 0$ is a saddle point of $S(1+z^4)$. Standing on this pass at the foot of four symmetrical ridges, separating the four valleys containing the four desired pits, and looking north, east, south, and west, the parachutist always finds the terrain higher than he is, no matter how short his proposed step. Hence he finally announces incorrectly that he is at the bottom of a pit.

Naturally a slight perturbation in his position or a tiny variation of his bearing from true north will lead our hero on a path to one of the true zeros of $1 + z^4$. In most practical applications such perturbations are always present in round-off noise which is one of the very few good features of this error. However in some simple exact situations, as the one above, there is no escape from the symmetry.

This "down hill" method which explores only four points in the neighborhood of a point P on $S(f)$ can be made much more sophisticated by allowing a larger number of possible directions as candidates for the next step. This gives us the general discretization of the method of steepest descent. Naturally it will still be possible for us to construct an example for which the method fails once we are told the angles at which the parachutist takes his bearings. However, the improved procedure, though more costly, leads more directly to a desired zero with less of the staggering and bouncing from side to side observed with only the four directions available.

In case of fog, the parachutist has another more sophisticated procedure available. For this we require the existence of the partial derivatives

$$D_x = \frac{\partial}{\partial x}||f(x)||, \quad D_y = \frac{\partial}{\partial y}||f(z)|| \qquad (z = x + iy).$$

Then the correct step Δz is that implied by the components

$$\Delta z = \Delta x + i\Delta y; \quad \Delta x = -hD_x, \quad \Delta y = -hD_y$$

whose direction is of course independent of h. The loss of altitude resulting from this step is approximately

$$(3) \qquad \Delta ||f(z)|| = -h(D_x^2 + D_y^2).$$

Here the parachutist is navigating by formula without having to take measurements of elevation.

Glancing at (3) one is tempted to allow him to choose not only his direction but also his value of h so as to reduce the number of steps to his goal. With this amount of freedom we now have a sort of Newton-Bairstow procedure with its beneficial quadratic convergence but with occasional disasters requiring supervision, or at least more or less built-in safeguards in case our man encounters wrinkled or bumpy regions of $S(f)$.

A few general comments can be made about the difficulties of locating a pit in the surface $S(f)$. In case of a simple zero of $f(z)$ the pit is nearly conical with a cross section whose shape is that of our norm's "unit circle" $||z|| = 1$.

For the class of ℓ_p norms

(4) $$||z||_p = \{|x|^p + |y|^p\}^{1/p} \qquad (0 < p \leq \infty)$$

of which the cheapest are

$$||z||_\infty = \max\{|x|, |y|\}$$

$$||z||_2 = |z|$$

$$||z||_1 = |x| + |y|$$

the unit circle is symmetric about $x = 0$, $y = 0$ and $x = \pm y$, and is square, circular and diamond shaped in the respective cases of $p = \infty$, 2, and 1.

For the norms (4) and for a fixed zero of f, the walls of the pit will be steeper the smaller p is taken; for a fixed norm they will be steeper the larger $||f'(z_0)||$ happens to be. (Of course at such conical pits the partial derivatives D_x and D_y do not exist.) A steep pit means a good determination of the desired simple zero of f.

For a double zero of f the pit is no longer conical but has a rounded bottom which makes the determination of this minimum much less certain. The higher the multiplicity of the zero the worse matters become. Tightly clustered roots can be very difficult to deal with automatically by the above methods, especially in the presence of round-off noise.

After looking at some possible features of $S(f)$, instead of attempting to extract and use *more* information from the immediate neighborhood of the point P, one may decide to use *less*. In this way, at least, one's chances of being mislead by the local regime at P are reduced. The very minimum of information about the neighborhood of P is the elevation $||f(P)||$ of the point P. If we use only this, we are driven to a more or less random sampling of points P in the hope of finding a zero by accident. Obviously this will not do. Nevertheless, the prospect of using a number of non-walking parachutists for each root is not without merit, provided they can give a more global account of the surface $S(f)$, especially as to the geometry of the zeros of f.

The theory of analytic functions and especially polynomials of a complex variable provides us with information allowing us to devise inexpensive rational algorithms for answering global questions about the presence or absence of zeros of a function in certain standard regions that need not be merely neighborhoods.

Armed with such an algorithm there remains the question of its practical application to produce an exhaustive search procedure, preferably without need of human intervention.

One instance of such a method was proposed by the writer in 1961 and, since it is still largely unknown, it may as well be described in the present context [1]. The basic algorithm here applies to polynomials and answers the question: Does a given polynomial f have a zero inside a given circle? The practical application of this algorithm to the search for a zero of $f(z)$ is now described.

If $f(0) = 0$ we have found a zero of f. If not, we now begin by asking whether f has a zero inside the unit circle. If so, we repeat the question about the circle $|z| = 1/2$; if not we repeat the question about $|z| = 2$. Continuing in this manner we soon find an annulus

$$R < |z| < 2R$$

containing at least one zero of f, the inner circle $|z| = R$ having no zero of f.

This annulus can be completely covered by eight equally spaced overlapping circles C_j of radius $4R/5$ whose centers are

$$c_j = \frac{3R}{2} \sec \frac{\pi}{8} e^{2\pi i j/8} \qquad (j = 0(1)7)$$

$$(\frac{3}{2} \sec \frac{\pi}{8} = 1.6235883).$$

At least one of these eight circles contains a zero of f. By applying our algorithm at most seven times, we can find the first such circle. This discovery will take place sooner, on the average, if we examine the circles not in increasing order of argument but in the order

$$j = (0,3,6,1,4,7,2)$$

or in the order

$$j = (0,4,2,6,1,3,5)$$

Having found an appropriate circle C_j with center c_j, we have completed the first step. The next step consists in treating C_j the way we treated the unit circle in case it covered a zero, as does C_j. That is, we find another annulus

$$R_1 < |z - c_j| < 2R_1$$

where

$$R_1 = (4/5)R \cdot 2^{-m} \quad m > 0$$

and begin to explore it with covering circles of radius

$$\frac{4}{5}R_1 \leq \frac{8}{25}R.$$

Having found one covering a zero we have completed step 2. After completing step k we have the center of a circle of radius less than $2R(2/5)^k$ (and probably much smaller) covering a zero. A couple of dozen steps at most will give the position of the zero to within an error less than $R \cdot 10^{-10}$.

Accepting this approximate zero z_1, or perhaps improving its value by one of the methods described earlier, we can now remove the zero:

$$f_1(z) = f(z)/(z - z_1)$$

and repeat the process with f_1, this time starting, not with the unit circle, but with the circle $|z| \leq R$. Finally all zeros, or a prescribed number of zeros, will be determined, approximately in order of increasing absolute value.

The algorithm for deciding whether a given circle covers a zero of $f(z)$ is rational in the coefficients of f and is based on an idea of I. Schur (1918). In the first place we may make a linear transformation on z to bring the given circle $|z - c| = r$ to the unit circle. Suppose under this change of variable, $f(z)$ becomes $g(z)$ with $g(0) \neq 0$. Let us write

$$f(rz + c) = g(z) = a_0 + a_1 z + \ldots + a_n z^n \qquad (a_0 a_n \neq 0)$$

where the coefficients will usually be complex. We define the associated polynomial $g^*(z)$ by

$$g^*(z) = \bar{a}_n + \bar{a}_{n-1} z + \ldots + \bar{a}_0 z^n .$$

Eliminating z^n between $g(z)$ and $g^*(z)$ gives us the polynomial of lower degree

$$T(g) = T(g(z)) = \bar{a}_0 g(z) - a_n g^*(z).$$

If $T(g(0)) = |a_0|^2 - |a_n|^2$ is negative then $g(z)$ has at least one zero inside $|z| = 1$ and so $f(z)$ has a root inside its given circle $|z - c| = r$. If $T(g(0))$ is positive we continue, repeating the process on $T(g(z))$ to obtain

$$T(T(g(z))) = T^2(g(z)).$$

If $T(g(0)) = 0$ we stop. In general we continue to generate polynomials

$$T(g(z)), \; T^2(g(z)), \; T^3(g(z)), \; \ldots$$

until we either encounter $T^h(g(z))$ with $T^h(g(0)) < 0$ or we obtain only positive values of

$$T(g(0)), \; T^2(g(0)), \; \ldots, \; T^{k-1}(g(0))$$

and $T^k(g(0)) = 0$. In the former case $f(z)$ has a zero inside its given circle. If in the latter case $T^{k-1}(g(z))$ is a constant, then $f(z)$ has no zero inside its circle.

In case $T^{k-1}(g(z))$ is not a constant the algorithm gives us no information. This happens for example when f has a zero on, but no zero inside, its given circle. In these cases we can restate our question with a new circle larger by a modest factor like $3/2$. This maneuver can also be executed whenever we encounter a value of $T^h(g(0))$ so small in absolute value that its sign is uncertain. In this way we are avoiding putting the circumference of our circle too close to a pit of $S(f)$.

Enough has been said to convince the enterprising programmer that he can build from these ideas a foolproof complex equation solving subroutine; and this is indeed the case. However the price for the insurance against disaster or convergence failure is not negligible. In case one has a sequence of equations of moderate degree with real coefficients and real distinct roots which lie in easily predictable intervals the classical methods are cheaper.

We can return to our analogy and say that we have just discussed in detail a search procedure in which our parachutist is equipped with a portable instrument to detect the presence of a pit in $S(f)$ at a given distance from where he is standing. Using a helicopter he can be transported from place to place depending on his reported observations. Using a little more theory, we can devise more expensive procedures corresponding to a more elaborate instrumentation of our flying surveyor. For example we can provide him with an instrument to count (with their multiplicities) the number of pits within a given distance or to measure the distance to the nearest pit. Each type of instrumentation has a suitable strategy for discovering a pit in minimum average time. As with our earlier discussion, the extra expenditure for additional information at a point of $S(f)$ does not also buy peace of mind.

Instead of using circles to contain or cover zeros one can use half planes. It is a pleasure to recall in this connection the well known contribution of one of Zurich's most illustrious mathematicians, Adolph Hurwitz. He gave in 1895 his criterion for a polynomial of degree n with real coefficients to have all its n zeros in the left half plane $Re(z) < 0$.

In one form, the criterion calls for a sequence of n determinants, formed from the coefficients, all to be positive. For $n = 4$ and

$$f(z) = z^4 + A_1 z^3 + A_2 z^2 + A_3 z + A_4$$

the relevant determinants are $D_1 = A_1$

$$D_2 = \begin{vmatrix} A_1 & A_3 \\ 1 & A_2 \end{vmatrix} \qquad D_3 = \begin{vmatrix} A_1 & A_3 & 0 \\ 1 & A_2 & A_4 \\ 0 & A_1 & A_3 \end{vmatrix}$$

$$D_4 = \begin{vmatrix} A_1 & A_3 & 0 & 0 \\ 1 & A_2 & A_4 & 0 \\ 0 & A_1 & A_3 & 0 \\ 0 & 1 & A_2 & A_4 \end{vmatrix}$$

In case the coefficients are complex the determinant D_k is to be replaced by a certain determinant of order $2k - 1$ in the real and imaginary parts of the coefficients.

In any case there is a rational algorithm involving only the coefficients of $f(z)$ for deciding whether $f(z)$ has any zeros to the left of the imaginary axis.

Once we have this criterion as a subroutine there are a number of ways in which it can be used, one being the following. Applying the criterion to $f(z)$ it may turn out that all the zeros of $f(z)$ lie in $Re(z) < 0$. If so, we can reapply it to $f(z-1)$, $f(z-2)$, $f(z-4)$ etc.; if not, we apply it to $f(z+1)$, $f(z+2)$, $f(z+4)$, etc. In any case we arrive at the half-plane $Re(z) < k$, containing all the zeros of $f(z)$. Changing z to $-z$, iz and $-iz$ and repeating this we can find a rectangle R parallel to the axes containing all the zeros of $f(z)$. We replace this rectangle by the square Q with the same center as R whose side is the larger of the sides of R. The four quadrants of Q can now be covered by their four circumscribing circles and using the previously described algorithm, we ask in turn whether these circles cover a zero of $f(z)$. More precisely, we may proceed as follows: If the first quadrant circle, in the north-east, covers a zero of f we replace Q by the square Q_1

circumscribing this circle with sides parallel to the axes. If this
circle fails to cover a zero of f we try the third quadrant circle
next. If there is success in this case we adopt for Q_1 the corresponding circumscribing square. If not we finally try the second quadrant circle, choosing the corresponding circumscribing square for
Q_1 in case of success. If there is still failure we take the fourth
quadrant of Q as Q_1. In any case we treat Q_1 in the way Q was
treated, obtaining by iteration a sequence of covering squares whose
areas are decreasing at least as rapidly as 2^{-m}. The process ceases
when the size of the square meets a specified tolerance which may
be a function of z. Dividing out the zero represented by the center
of the last square Q_N from $f(z)$ (as previously) we return to the
square Q and begin the iteration again in search of a new zero of
f.

Comparing this method with the one using eight circles to
cover an annulus one finds that it is indeed competitive. The convergence rates are $1/\sqrt{2}$ and $2/5$, respectively, but the average
numbers of circles used are 2 and 4, respectively, at each step.
The relative costs of the two have therefore as a ratio

$$\frac{log5 - log2}{log2} = 4/3.$$

That is to say, the "circle method" is about 33 percent cheaper than
the "square method". Perhaps the fact that the circle method brings
out the roots in increasing order of size is a more significant
advantage than the slightly lower cost.

In conclusion there are a few general remarks that can be
made about search methods.

One general principle to attempt to follow is this: Do not ask embarrassing questions about the environment if you wish to avoid instability; it only makes the algorithm "nervous". For example do not attempt, in the square method, to obtain the smallest rectangle enclosing all the zeros of f. If, in the circle method, a root lies very near to the circumference of a large testing circle, either inside or outside, there will be a nervous reaction by the algorithm resulting in a small number whose sign must be determined. In this case it is not worthwhile forcing the issue. It is more diplomatic to rephrase the question using a somewhat larger circle.

The well known injunction: "Don't divide by zero" can be strengthened to read "Don't divide". It will be observed that there is no instance of the use of this unstable operator in any of the proposed search methods. Subtraction is unstable enough.

There is a need to devote some thought to the comparison of different methods of polynomial equation solving. It is easy to compare convergence rates and expected average costs of carrying out the calculations in two different applications to the same polynomial. What is more uncertain is the assessment of the goodness of fit of the answers.

REFERENCES

1. D.H. Lehmer, "A Machine Method for Solving Polynomial Equations", Jour.Assoc.Comp.Machinery, 8 (1961), 131-162.

2. J.A. Ward, "The down hill method of solving $f(z) = 0$", Jour. Assoc.Comp.Machinery, 4 (1957) 148-150.

Prof. D.H. Lehmer
Department of Mathematics
University of California
Berkeley, Cal. 94720

A. M. Ostrowski

A Method for Automatic Solution of Algebraic Equations*

1. Cauchy's famous existence proof for the roots of Algebraic Equations is based on the fact that, given a polynomial

$$f(z) = A_0 z^n + A_1 z^{n-1} + \ldots + A_n,$$

to any z_0 such that $f(z_0) \neq 0$, there exists a $z = z_0 + h$ such that

$$|f(z)| < |f(z_0)|.$$

To prove this, Cauchy develops $f(z)$ in powers of h,

$$f(z) = \sum_{\nu=0}^{n} a_\nu h^\nu$$

and writes it, as $a_0 = f(z_0) \neq 0$,

---

*) This method was developed in part under the Contract of US Army with the Mathematical Institute of the University of Basel, in part at Mathematics Research Center, US Army, Madison, Wisconsin, jointly with V. Pereyra, and in part at the IBM, T.J. Watson Research Center, Yorktown Heights, New York.

$$f(z) = a_0 + a_k h^k + a_{k+1} h^{k+1} + \ldots = a_k \left[\frac{a_0}{a_k} + h^k (1 + \varepsilon(h)) \right],$$

$$\varepsilon(h) = b_1 h + b_2 h^2 + \ldots .$$

Here we assume that $a_1 = \ldots = a_{k-1} = 0$, $a_k \neq 0$. If $a_1 \neq 0$, then $k=1$. If we now set

$$h = \rho \sqrt[k]{-\frac{a_0}{a_k}}, \quad \rho > 0,$$

we obtain

$$f(z) = a_0 \left[1 - \rho^k (1 + \varepsilon(h)) \right],$$

and, as $\varepsilon(h) \to 0$ with $\rho \to 0$, the modulus of the bracketed expression can, for sufficiently small ρ, be made $< 1 - \rho^k/2 < 1$.

2. Observe that normally we will have $a_1 = f'(z_0) \neq 0$, $k=1$, so that h is given by

(2.1) $$h = -\rho \frac{a_0}{a_1} = -\rho \frac{f(z_0)}{f'(z_0)},$$

and we see that the direction from z_0 to z is in this case given by Newton's quotient $-f(z_0)/f'(z_0)$. However, the direct estimate of ρ from the development of $f(z)$ at z_0 requires a rather costly computation and gives a very small value of ρ. Further, the procedure indicated in the case $k > 1$ has little sense from the point of view of numerical analysis, because it is based on the *exact vanishing* of the intermediate coefficients $a_1, \ldots, a_{k-1}$. On the other hand, some similar procedure is necessary, since the iteration in the

Newton direction gives sometimes a convergent sequence which converges to a zero of $f'(z)$ instead of $f(z)$.

3. In order to obtain from Cauchy's idea a workable algorithm, we have therefore to solve the following problems:

(a) To obtain, if $f'(z) \neq 0$, a good estimate for ρ, which does not depend on the complete Taylor development of $f(z)$.

(b) To find a convenient method for accelerating the sequence of iterations obtained using (a).

(c) To obtain a test for z_0 being close to a zero of $f'(z)$ distinct from zeros of $f(z)$, and a routine for skipping such a zero in order to eliminate convergence to zeros of $f'(z)$.

(d) A test to recognize whether z_0 is already sufficiently close to a simple zero of $f(z)$ so that the classical Newton-Raphson routine can be used.

(e) A test to recognize whether z_0 is sufficiently close to a cluster of m zeros of $f(z)$, and to obtain in this way a bound for the error committed, taking z_0 as an approximate root and m as an "approximate multiplicity" of z_0.

4. Before attacking the problem, the polynomial $f(z)$ is reduced to the form

(4.1) $$f(z) = z^n + a_2 z^{n-2} + \ldots + a_n, \quad \max_{2 \leq \nu \leq n} |a_\nu| = 1.$$

Then it is easily seen that we have

(4.2) $$|f(z)| > 1 \quad (|z| > 2)$$

and all zeros ζ of $f(z)$ satisfy the inequality

(4.3) $$|\zeta| \leq \frac{1 + \sqrt{5}}{2} = 1.62\ldots\ .$$

This transformation is important for the following reason:

In our procedure, for any approximation z_ν, the next approximation $z_{\nu+1}$ is only accepted if we have

(4.4) $$|f(z_{\nu+1})| < |f(z_\nu)|\ .$$

Therefore, if we begin with z_0, such that $|f(z_0)| \leq 1$ (for instance $z_0 = 0$), we are sure that the following z_ν will never leave the circular disc $|z| \leq 2$. On the other hand, for a polynomial of the type (4.1), we can use universal bounds for $|f'(z)|$, $|f''(z)|,\ldots$, in $|z| \leq 2$, depending only on n.

5. The problem (a) can be solved in different ways. We discuss two of them.

I. From the theory of the method of steepest descent, it follows that by setting

$$\Lambda^* = 2 \max_{|z| \leq 2} (|f'|^2 + |ff''|)$$

we obtain for

(5.1) $$z = z_0 - \frac{f(z_0)\bar{f}'(z_0)}{\Lambda^*}$$

the relation

$$(5.2) \qquad |f(z_0)|^2 - |f(z)|^2 \geq 2 \frac{|f(z_0)f'(z_0)|^2}{\Lambda^*} .$$

Hence, if we iterate according to (5.1), the obtained sequence z_ν is certainly convergent either to a zero of $f(z)$ or to a zero of $f'(z)$.

II. The direct approach to the problem (a) is based on the following inequality (5.5): Let

$$(5.3) \qquad R(z) = \frac{f(z)}{f'(z)} .$$

Assuming that $R(z) \neq 0, \neq \infty$, set

$$(5.4) \qquad T(z) = \frac{|f'(z)|}{M_2|R(z)|} , \quad M_2 = \max_{|z|\leq 2} |f''(z)| .$$

We have then for any real or complex t, as long as $|z-tR(z)| \leq 2$,

$$(5.5) \qquad \left|\frac{f(z-tR(z))}{f(z)}\right| \leq |1-t| + \frac{t^2}{2T} .$$

Hence, for $t=T$ and $t=1$:

$$\left|\frac{f(z - TR)}{f(z)}\right| \leq 1 - \frac{T}{2} \quad (T \leq 1),$$

$$\left|\frac{f(z-R)}{f(z)}\right| \leq \frac{1}{2T} \leq \frac{1}{2} \quad (T \geq 1),$$

so that ρ in (2.1) can be taken as

$$(5.6) \qquad \rho_d = \min\left(1, \frac{|f'|^2}{M_2|f|}\right) .$$

If we compare ρ_d with the value ρ_g, obtained from the method of steepest descent,

$$(5.7) \qquad \rho_g = \frac{|f'|^2}{max(|f'|^2 + |ff''|)} ,$$

we see that $\rho_d > \rho_g$. Furthermore, in computing $T(z)$, we recognize very easily the neighborhood of a simple zero of $f(z)$ and the applicability of the Newton-Raphson procedure.

However, the flow chart given in the appendix corresponds to the choice of ρ_g and to the iteration procedure arising from the method of steepest descent. We will give in another report the discussion corresponding to the choice $\rho = \rho_d$.

6. We now come to problem (b). We estimate the computational work mostly in terms of *Horner units*. A *Horner H* is the amount of computational work necessary for computing the value of a polynomial $a_0 z^n + a_1 z^{n-1} + \ldots + a_n$. For a polynomial of degree n a Horner reduces to $n(M+A)$ where M denotes a multiplication and A an addition. Of course, the exact amount of machine time needed for an average M or A depends on the type of the machine and also on the type of arithmetic used, whether it is real or complex, single- or double- or even multiple-precision arithmetic.

7. We first used, in [5], the Steffensen accelerating procedure. Subsequently in [4], I introduced yet another method.

Take a fixed $q > 1$ and set

$$(7.1) \qquad w(z) = \frac{f(z)\overline{f}'(z)}{\Lambda^*} ,$$

(7.2) $$\varphi(S) = |f(z - Sw(z))|.$$

We know that we have $\varphi(1) < \varphi(0)$.

Take the smallest integer $k = k_q(z)$, so that

(7.3)
$$\varphi(1) > \varphi(q) > \varphi(q^2) > \ldots > \varphi(q^k),$$
$$\varphi(q^k) \leq \varphi(q^{k+1}).$$

Then set

(7.4) $$G^*(z) = z - q^k w(z)$$

and consider the iteration

(7.5) $$z_{\nu+1} = G^*(z_\nu).$$

It can then be proved that

(7.6) $$\overline{\lim} \left| \frac{G^*(z)-\zeta}{z-\zeta} \right| \leq \frac{q-1}{q+1} \qquad (z \to \zeta),$$

if z tends to a (simple or multiple) zero of $f(z)$.

8. (7.6) shows that the rate of convergence of our iteration is faster for small values of q, while, on the other hand, for large values of q the computation of $k_q(z)$ requires considerably less time.

In our method, we try to combine both possibilities, using first $q = 8$ and then checking again with $q = 2$. In this way, a considerable acceleration can already be obtained. Theoretically, an even better acceleration could be obtained, reducing the number of comparison trials from k to $O(\log k)$, if we search for k, representing it as a sum of Fibonacci numbers. From the programmer's point of view, however, this presents a certain complication, since provision must be made for storing the Fibonacci numbers up to the greatest used.

9. On the other hand, after the optimal value of 2^k has been obtained, we can improve the result further by searching in the neighborhood of 2^k for an even more convenient integer S, which does not need to be a power of 2. This can be done, applying the standard procedure for weighting with a diadic number by approximating the weight alternatively from above and from below. The mathematical theorem behind this procedure can be formulated [1] as the possibility of the unique development of any natural integer S in the form:

$$S = 2^{m_1} - 2^{m_2} + 2^{m_3} - \ldots + 2^{m_{2S+1}}, \; (m_1 > m_2 > \ldots > m_{2S+1} \geq 0) \quad *)$$

---

*) Instead of using the diadic development, we could also use the triadic development, which offers a slight advantage. Since, however, the acceleration obtained in this way is, on the average, only around 2 % but, on the other hand, programming becomes technically somewhat more complicated, the method using the diadic development appears to be preferable.

10. To discuss problem (c), replace z_0 by z and develop $f(z+h)$ at z,

(10.1) $\quad f(z+h) = \sum_{\nu=0}^{n} D_\nu h^\nu, \quad |D_\nu| = \delta_\nu \quad (\nu = 0,1,\ldots,n)$.

Then, if we are close to a cluster of m roots of $f'(z)$, and not in a neighborhood of a zero of $f(z)$, we must express conveniently that $\delta_1,\ldots,\delta_m$ are relatively small, while δ_0 and δ_{m+1} are not too small.

We have to introduce

(10.2) $\quad \kappa = 3^{-\frac{1}{m+1}}$,

$$\mathbf{M}_m = \delta_{m+2} + \frac{1}{10}\delta_{m+3} + \frac{1}{10^2}\delta_{m+4} + \cdots .$$

If then we set

(10.3) $\quad L_m = \min\left(\frac{1}{10}, \left(\frac{\delta_0}{\delta_{m+1}}\right)^{\frac{1}{m+1}}, \frac{\delta_{m+1}}{4\mathbf{M}_m}\right)$,

our condition can be expressed as ([2], pp. 6-7)

(10.4) $\quad \max_{1 \leq \nu \leq m} \frac{1}{\kappa}\left(\frac{3m\delta_\nu}{\delta_{m+1}}\right)^{\frac{1}{m+1-\nu}} \leq L_m, \quad |z| \leq \frac{19}{10}$.

Checking whether (10.4) is satisfied for a suitable m is the J Routine.

If (10.4) is satisfied, then it can be proved ([2], Theorem 2) that, taking $\rho = L_m$, there are exactly m zeros $\zeta'_1,\ldots,\zeta'_m$ of $f'(w)$ in the ρ-neighborhood of z. Setting then

(10.5)
$$\vartheta = \frac{1}{m+1}(\arg f(z) - \arg f^{(m+1)}(z) + \pi),$$
$$z^* = z + e^{i\vartheta}\rho \equiv J(z),$$

we prove that we have

(10.6)
$$|f(z^*)| < |f(\zeta'_\mu)| \quad (\mu = 1,\ldots,m),$$
$$|f(z^*)| < |f(z)|.$$

(The J Routine).

It follows then from (10.6) that, if we restart our iteration from z^*, the obtained sequence cannot converge to one of the roots $\zeta'_1,\ldots,\zeta'_m$ of $f'(z)$.

11. The J Test is relatively expensive in time, particularly as we have to apply it for all values of $m=1,\ldots,n-1$. Since, however, the probability of $m > 1$ is very small, we use the complete J Test only for every tenth approximation and, for intermediate approximations, restrict ourselves to the test J_1, corresponding to $m = 1$. For $m = 1$, we can use even better constants than obtained from (10.3) - (10.4), setting there $m = 1$ ([2], p. 10).

The J_1 Test consists in checking the inequality:

(11.1)
$$\frac{3\delta_1}{\delta_2} \leq \rho \equiv \min\left(\frac{1}{10}\sqrt{\frac{\delta_0}{\delta_2}}, \frac{\delta_2}{2M_3}\right),$$

where $M_3 = \max_{|z|\leq 2} \frac{|f'''(z)|}{6}$, $|z| \leq \frac{19}{10}$.

If the J_1 Test is satisfied, we use the J Routine with $m = 1$ and with a value ρ from (11.1).

12. As to problem (d), the routine tests can be used. See for instance [3].

However, the overall bound for M_2 being sometimes too large, some improvements are possible and useful, which we will not discuss here.

13. To formulate a test corresponding to problem (e), we must express in a convenient way that $\delta_0, \ldots, \delta_{m-1}$ are small as compared to δ_m. The exact formulation of the Ω Test, which we use for this purpose, is:

(13.1)
$$\rho \equiv \min\left(\frac{1}{10}, \frac{\delta_m}{2\mathbf{M}_{m-1}}\right) > \max_{0 \leq \mu \leq m-1} \left(\frac{2m\delta_\mu}{\delta_m}\right)^{\frac{1}{m-\mu}} \equiv r,$$
$$|z| \leq \frac{19}{10}.$$

If (13.1) is satisfied, then we have exactly m zeros of $f(z)$ in the ρ-neighborhood of z, and they lie even in the r-neighborhood of z ([2], Theorem 1).

The Ω Test is again pretty expensive, and therefore we use mostly the test corresponding to $m = 1$, namely

(13.2)
$$\delta_0 < \min\left(\frac{\delta_1}{6}, \frac{\delta_1^2}{n^2 2^n}\right), \quad |z| \leq \frac{5}{3}.$$

If this is satisfied, then we have exactly one zero in the ρ-neighborhood of z, $\rho = \frac{\delta_1}{n^2 2^n}$, and this zero lies even in the r-neighborhood of z, $r = \frac{2\delta_0}{\delta_1}$.

14. After one root of the equation has been obtained, the corresponding linear factor is divided out of $f(z)$ by Ruffini-Horner division. If the equation is real and the root obtained non real, the corresponding quadratic factor is divided out. Then the same procedure is applied to the reduced polynomial obtained in this way, and so on. The reader may consult the flow charts given in the Appendix.

15. To what extent can the above procedure fail? First of all, since the value of Λ^* given above is very large for large n, it could easily happen that the value of $w(z)$ cannot be distinguished by the computer from 0. In this case, we replace $w(z)$ by $2^p w(z)$, taking the smallest positive integer p for which $2^p w(z)$ can be distinguished by the computer from 0, and check whether or not

(15.1) $$|f(z - 2^p w(z))| < |f(z)| .$$

In the first case (and this will in general occur for $n \leq 10$), we replace in the definition of $\varphi(S)$ in (7.2) $w(z)$ by $2^p w(z)$ and proceed as in sections 7-9. In the second case, if the Ω Test does not show that z is already a sufficiently good approximation to a root of $f = 0$, single precision is obviously inadequate.

16. A similar situation arises if 2^k turns out to be too large for the machine. Since in any case $2^k |w(z)| < 4$, here again $w(z)$ is too small, and we have to replace $w(z)$ by a convenient multiple $2^p w(z)$. In this case, (15.1) is obviously satisfied.

17. The most important reason for failure of the above method, used directly, is of course the accumulation of round-off errors. This accumulation is particularly dangerous in the case of multiple roots or clusters of roots. It is, of course, not difficult to build into the above procedure provisions for round-off errors from pretty rough estimates up to the complete treatment by Interval Analysis.

18. However, as in such cases the necessity of using at least double precision arises rather often, we are of the opinion that it is better to prepare two programs for the above method.

The first "simplified program" would proceed by simple precision without accounting for round-off (except in the application of the Ω Test, in order to secure exact bounds) provided it does not fail.

If on the other hand, the simplified program fails, then the second "complete program" would have to be used, which has to be written from the beginning in double precision and must account for round-off errors in the most complete way possible, that is by using Interval Analysis. Of course, only after sufficient experience has been gathered with the simplified program, it can be decided whether some further provisions for round-off ought to be introduced into the simplified program itself.*)

---

*) A program, corresponding to our method, has been prepared at the IBM Center, Yorktown Heights, and many examples computed with this program appear to show that our method is indeed completely within the range of practical computation. However, these examples do not give a sufficient basis for deciding the merits of the simplified and the complete programs, since the programmer who prepared the program incorporated some features without discussing them with the author, so that we cannot assume any responsibility for this program.

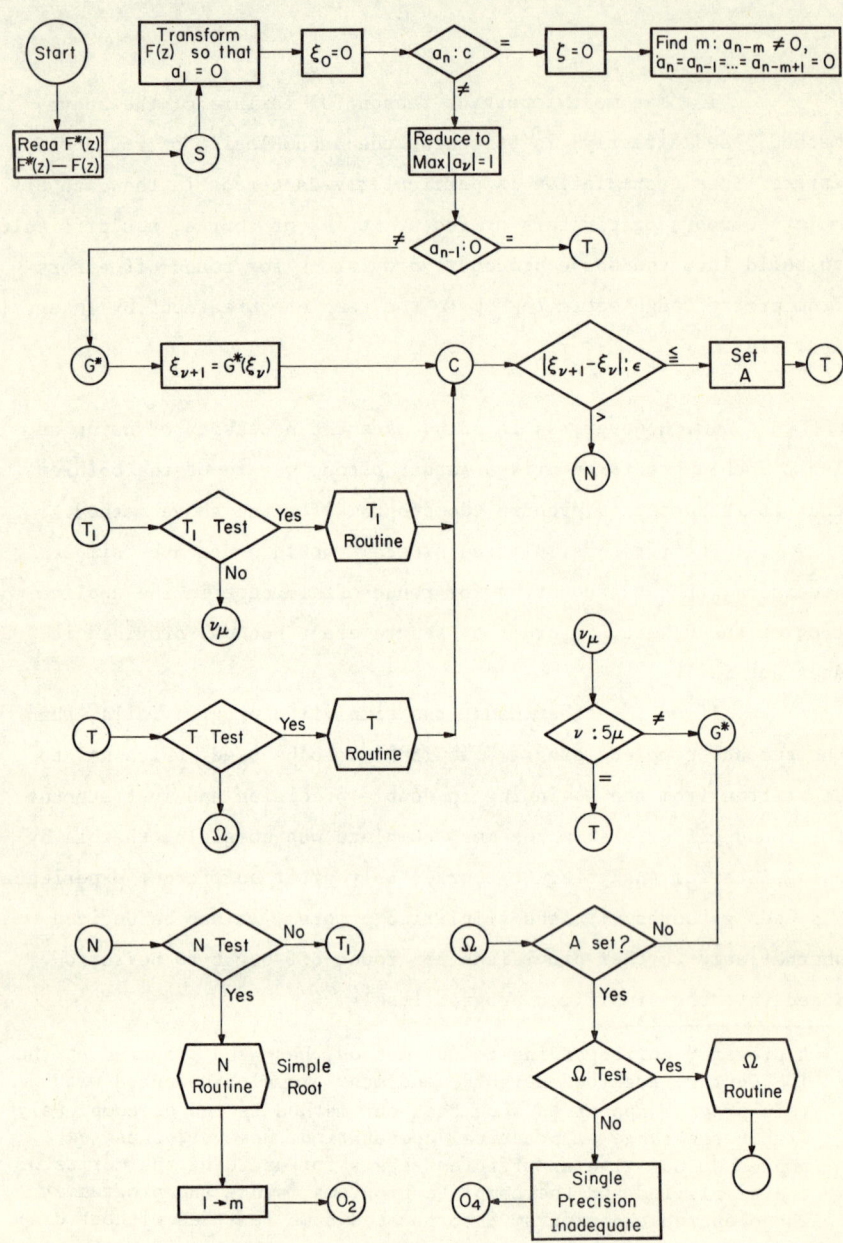

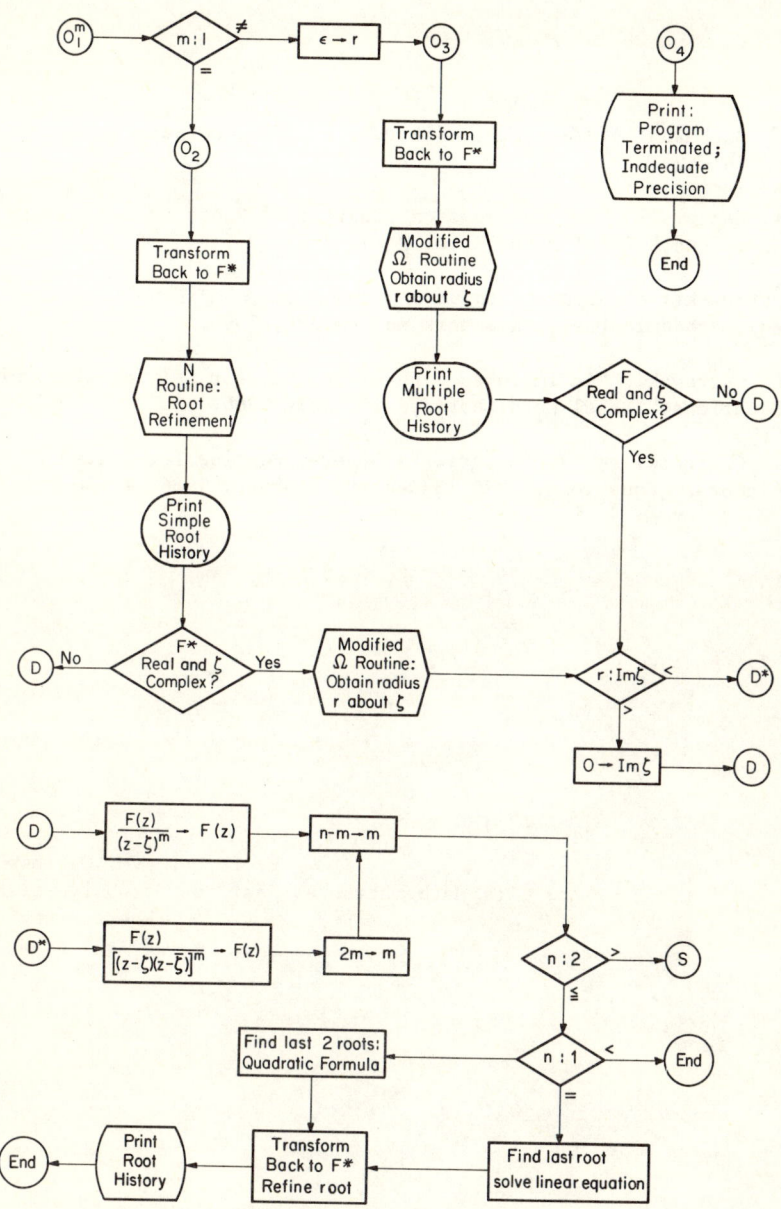

REFERENCES

1. P. Bachmann: *Niedere Zahlentheorie*, 2. Teil, Leipzig 1910, pp. 61-62.

2. A. Ostrowski: "Two Theorems on Clusters of Roots of Polynomial Equations," Basel Math. Notes # 17, June 1967.

3. Ostrowski: *Solution of Equations and Systems of Equations*, 2nd ed., Academic Press, New York and London, 1966.

4. A. Ostrowski: "An Improved General Routine for Solving Algebraic Equations," Basel Math. Notes, # 15, July 1966.

5. A. Ostrowski and V. Pereyra, "A General Routine for Solving Algebraic Equations," MRC TS Report, February 1966, # 630.

Prof. Dr. A.M. Ostrowski
Mathematisches Institut der Universität
Rheinsprung 21
Basel, Schweiz

Monica Pavel-Parvu and André Korganoff

Iteration Functions for Solving Polynomial Matrix Equations

CONTENTS

1. Introduction	226
2. Classification of iteration functions	227
3. Convergence. Ordinary iteration functions	230
3.1 - Non-singularity of the linear operator	230
a) - Linear convergence	232
b) - Superlinear convergence	234
3.2 - Singularity of the linear operator	236
a) - Linear convergence	243
b) - Superlinear convergence	246
3.3 - Nullity of the linear operator at the solution	247
3.4 - Multiple roots	250
3.5 - Commutativity	252
4. Convergence. General iteration functions	255
5. Iteration functions	256
5.1 - General theorems	260
5.2 - Iteration functions of Newton and Schröder	262
5.3 - Other iteration functions	265
5.4 - Multiple roots	267
Appendix 1. Fréchet differentials	269
2. Convergence of iterative methods for solving singular linear systems by matrix splitting	275
References	279

1. Introduction

One knows theoretically how to solve a certain number of polynomial matrix equations. Let us mention among the most classical ones the unilateral polynomial equations (the results concerning these can besides be extended - see F.R. Gantmacher [1], p. 288 - to rectangular coefficient matrices) and the more particular equation $x^{\nu} = a$ which gives the νth root of a. Let us also mention the quadratic equation

(1) $$a + bx + xb^* - x\,c\,x = 0,$$

which occurs in the search for stable solutions of the differential matrix equation of Riccati, and in which control theory is interested (see particularly W.T. Reid [2], [3], and J.E. Potter [4]). In any case, the effective calculation of solutions leads to an extremely expensive, if not prohibitive work.

In the case of the general polynomial equation

(2) $$p_{\nu}(x) = a_0 + \sum_{k=1}^{\nu} \sum_{\ell(k)} a_{k,1}^{\ell(k)}\, x\, a_{k,2}^{\ell(k)}\, x \ldots x\, a_{k,k+1}^{\ell(k)} = 0,$$

where $a_{k,j}^{\ell(k)}$ and x are square or rectangular matrices, one does not even possess the few things one had before. We have no information upon solutions, their discrete or continuous character, their shape, or their localization. They are not necessarily factorizable, and if we know some of them, we do not know how to eliminate them.

All these not very propitious circumstances being taken together, it seems that one could easily try and that it would be interesting to proceed by methods of approximation where initial values would vary. This can be satisfactory in certain cases, if for instance one has already some information about the solution (s) desired (approximate value or localization, uniqueness of the solution - for instance for the positive square root of a positive operator), or when the search restricts itself to finding an arbitrary solution having given characteristics (a hermitian solution of (1) if a and c are hermitian). In any other case, all information being excluded, it would mean proceeding blindly, with the possibility only, but not the certitude of obtaining some results.

Besides, these approximate methods already exist. It is enough to consider a matrix equation as a system of scalar equations. But one loses information since one breaks a structure. There is then good reason to maintain the matrix form and to work out a theory of iterative methods of solution. This is what we try to do in the sequel, the scalar model serving very naturally as our inspiration (see in particular J.F. Traub [5]).

2. Classification of Iteration Functions

Taking the non-commutativity of matrix operations into account, we can imagine the most general form of iteration functions. It will be more succinct if based on generalizations of scalar iteration functions, which are easily accessible, as for instance the one for Newton's method. The one-point iteration functions would thus be

$$\sum_{k_1} \alpha_{1,k_1}(x_i, g_1(x_i), \ldots, g_p(x_i)) \, x_{i+1} \beta_{1,k_1}(x_i, g_1(x_i), \ldots, g_p(x_i))$$

$$= \varphi_1(x_i, g_1(x_i), \ldots, g_p(x_i)),$$

$$\sum_{k_2} \alpha_{2,k_2}(x_i, g_2(x_i), \ldots, g_p(x_i)) g_1(x_i) \beta_{2,k_2}(x_i, g_2(x_i), \ldots, g_p(x_i))$$
$$= \varphi_2(x_i, g_2(x_i), \ldots, g_p(x_i)),$$

(3) ———————————————————————————————

$$\sum_{k_{p+1}} \alpha_{p+1,k_{p+1}}(x_i) g_p(x_i) \beta_{p+1,k_{p+1}}(x_i)$$
$$= \varphi_{p+1}(x_i).$$

The left term of the first equation depends linearly upon x_{i+1}; the functions α_1, β_1 and φ_1 depend at one and the same time explicitly and implicitly upon x_i; the implicit functions $g(x_i)$ are successively given by a linear relation, as in a triangularization process.

The multipoint iteration functions would have an analogous form, namely

$$\sum_{k_1} \alpha_{1,k_1}(x_i, x_{i-1}, \ldots, x_{i-q}, g_1(x_i, x_{i-1}, \ldots, x_{i-q}), \ldots, g_p(x_i, x_{i-1},$$
$$\ldots, x_{i-q})) x_{i+1} \beta_{1,k_1}(\text{———————————————————})$$
$$= \varphi_1(\text{———————————————————————}),$$
———————————————————————————————.

In the sequel we will consider only one-point iteration functions.

We will define ordinary iterations, that do not involve implicit functions $g(x)$, i.e.

(4) $$\sum_k \alpha_k(x_i) \, x_{i+1} \, \beta_k(x_i) = \varphi(x_i),$$

thus distinguishing them from general iterations (3). This is a minor distinction in the sense that in a convergence study the first relation (3) plays a leading part. It is the only one which contains a term in x_{i+1}; the functions $\alpha_1, \beta_1, \varphi_1$ depend in other respects upon x_i and their explicitation is not necessarily asked for in this case. This distinction is nevertheless necessary, as certain conditions must be set in certain circumstances on the functions $g(x)$.

We often have

(5) $$\varphi(x_i) = \sum_k \alpha_k(x_i) \, x_i \beta_k(x_i) + \psi(x_i),$$

with the result that (4) can also be written

(6) $$\sum_k \alpha_k(x_i) \, (x_{i+1} - x_i) \, \beta_k(x_i) = \psi(x_i).$$

These are seemingly equivalent iteration functions. We will see that, as a matter of fact, they are different when the linear operator defined by means of the left-hand terms is singular.

Let us set, with (4), (5) and (6):

(4') $$\Phi(x,y) = \sum_k \alpha_k(x) \, y \beta_k(x) - \varphi(x),$$

(5') $$\Psi(x,y) = \sum_k \alpha_k(x) \, y\beta_k(x) - \sum_k \alpha_k(x) \, x\beta_k(x) - \psi(x),$$

(6') $$\overline{\Psi}(x,y-x) = \sum_k \alpha_k(x) \, (y-x) \, \beta_k(x) - \psi(x).$$

Assuming that the sequence of iteration functions converges, the limit will be a solution of $\Phi(x,x) = 0$, $\Psi(x,x) = 0$ or $\psi(x) = 0$, $\overline{\Psi}(x,0) = 0$, or also $\psi(x) = 0$. Putting it differently, these are fixed points of the iteration functions under consideration. $\psi(x)$ is often a function depending upon $f(x)$ as well as its derivatives, where $f(x) = 0$ is the equation we have to solve. $\psi(x)$ vanishes together with f at the solution.

We will try to specify in the following sections these fixed points as well as convergence conditions for the iteration functions.

3. Convergence. Ordinary Iteration Functions

3.1 Non-singularity of the linear operator

Let us consider the iteration (4). It is a linear equation in x_{i+1}; let

(7) $$A_i = A(x_i) = \sum_k \alpha_k(x_i) \otimes \beta_k^T(x_i),$$
$$\tilde{\varphi}_i = \tilde{\varphi}(x_i) = [\varphi_1, \; \varphi_2, \ldots]^T,$$
$$\tilde{x}_{i+1} = [(x_{i+1})_1, \; (x_{i+1})_2, \ldots]^T,$$

where ⊗ denotes the tensor product. The solution of (4) is then given by

$$(8) \qquad A_i \tilde{x}_{i+1} = \tilde{\varphi}_i .$$

We will suppose that the successive matrices A_i are non-singular, and - as the sequence of the iterates $\tilde{x}_i$ converges to a value $\tilde{x}$ - that the matrix $A = A(x)$ is also non-singular. We thus have for any i

$$(9) \qquad \tilde{x}_{i+1} = A_i^{-1} \tilde{\varphi}_i ,$$

and

$$(9') \qquad \tilde{x} = A^{-1} \tilde{\varphi}$$

is a solution of

$$(8') \qquad A\tilde{x} - \tilde{\varphi} = 0 .$$

It follows that x is a solution of

$$(8'') \qquad \Phi(x,x) = 0,$$

and is a fixed point of the iteration. If we had taken (5) and (6), $\tilde{x}$ would have been a solution of $\tilde{\psi} = 0$ and x a solution of $\psi(x) = 0$.

Let us examine under which conditions convergence will be attained. Let therefore

$$(10) \qquad \varepsilon_i = x - x_i .$$

As $\Phi(x,y)$ is linear in y, the iteration (4) may be written:

(11) $\quad \Phi(x_i, x_{i+1}) = \Phi(x-\varepsilon_i, x) - \Phi'_y(x-\varepsilon_i, x; \Delta y = \varepsilon_{i+1}) = 0,$

where Φ'_y is the Fréchet derivative of Φ with regard to y (see Appendix 1); it is independent of y, and satisfies - taking notations (7) into account - the following relation:

(12) $\quad \tilde{\Phi}'_y(x_i, x; \Delta y = \varepsilon_{i+1}) = A_i \, \tilde{\varepsilon}_{i+1}.$

If $\Phi(x,y)$ is a polynomial in x of degree ν for instance, its Taylor series expansion will be finite, and we will have:

(13) $\Phi(x-\varepsilon_i, x) = \Phi'_x(x,x; \Delta x = -\varepsilon_i) + \dfrac{1}{2!} \Phi''_{xx}(x,x; \Delta x = -\varepsilon_i, \Delta x = -\varepsilon_i)$

$\qquad + \ldots + \dfrac{1}{\nu!} \Phi^{(\nu)}_{\underbrace{x \ldots x}_{\nu \text{ times}}} (x,x; \underbrace{\Delta x = -\varepsilon_i, \ldots, \Delta x = -\varepsilon_i}_{\nu \text{ times}}).$

a) **L i n e a r   c o n v e r g e n c e**

We will assume that

(14) $\qquad\qquad \Phi'_x(x,x; \Delta x) \neq 0,$

i.e.

(14) $\quad \sum_k [\alpha'_k(x; \Delta x) \, x\beta_k(x) + \alpha_k(x) \, x\beta'_k(x; \Delta x)] - \varphi'(x; \Delta x) \neq 0.$

Let

(15)
$$\begin{cases} \tilde{\Phi}'_x(x,x;\ \Delta x = -\varepsilon_i) = -B\,\tilde{\varepsilon}_i \,, \\[4pt] \dfrac{1}{2!}\,\Phi''_{xx} + \ldots + \dfrac{1}{\nu!}\,\Phi^{(\nu)}_{x\ldots x} = R \,, \\[4pt] \tilde{R} = -b_i\,\tilde{\varepsilon}_i \,, \\[4pt] B + b_i = B_i \,, \end{cases}$$

with b_i converging to zero, and B_i to B, as ε_i or $\tilde{\varepsilon}_i$ converge to zero.

(11) can thus be written:

(16) $$A_i\,\tilde{\varepsilon}_{i+1} = -B_i\,\tilde{\varepsilon}_i \,,$$

which yields

(17) $$\tilde{\varepsilon}_{i+1} = -A_i^{-1} B_i\,\tilde{\varepsilon}_i.$$

If for any i, the moduli of the eigenvalues of $A_i^{-1} B_i$ are uniformly smaller than one, $\tilde{\varepsilon}_i$ will tend to zero, and convergence will be achieved. $\tilde{x}_i$ will converge to $\tilde{x}$, and therefore x_i to x, A_i and B_i to A and B. $A_i^{-1} B_i$ being a continuous function, it suffices, if $\lambda(A^{-1}B)$ is an eigenvalue of $A^{-1}B$, that

(18) $$|\lambda(A^{-1}B)|_{max} \leq K < 1 \,.$$

(There does indeed exist a neighborhood of x, hence an η, such that, for every x_i within $||x - x_i|| \leq \eta$ - where $||.||$ is an arbitrary usual norm - one has $|\lambda(A_i^{-1} B_i)|_{max} \leq L < 1$).

If we had started with the iterations (5) and (6), we would let

(19)
$$\tilde{\psi}'_x(x,x; \Delta x = -\varepsilon_i) = \tilde{\psi}'_x(x,0; \Delta x = -\varepsilon_i)$$
$$+ \tilde{\psi}'_{y-x}(x,0; \Delta(y-x) = +\varepsilon_i) = (A + C)\tilde{\varepsilon}_i,$$

and we would have similar to above

(17')
$$\tilde{\varepsilon}_{i+1} = A_i^{-1}(A_i + C_i)\tilde{\varepsilon}_i,$$

and

(18')
$$|\lambda(I + A^{-1}C)|_{max} \leq K < 1.$$

We have assumed that all matrices A_i are regular. In practice it is possible to encounter a singular A_i. Anyway, A being regular, and being a continuous function in x, there exists a neighborhood of x where all matrices A are regular. We will then start with another value of x_i.

b) S u p e r l i n e a r   c o n v e r g e n c e

Let us assume that

(20)
$$\phi'_x(x,x;\Delta x) \equiv 0, \ldots, \phi^{(r-1)}_{x\ldots x}(x,x;\Delta x, \ldots, \Delta x) \equiv 0,$$
$$\phi^{(r)}_{x\ldots x}(x,x;\Delta x, \ldots, \Delta x) \neq 0.$$

Thus equation (11) becomes

$$\Phi'_y(x-\varepsilon_i, x; \Delta y = \varepsilon_{i+1}) = \frac{1}{r!} \Phi^{(r)}_{x\ldots x}(x, x; \Delta x = -\varepsilon_i, \ldots, \Delta x = -\varepsilon_i)$$
(21)
$$+ R(\varepsilon_i),$$

where $R(\varepsilon_i)$ is obtained from (13). $\Phi^{(r)}_{x\ldots x}$ is homogeneous and of degree r in Δx, such that it satisfies

(22)
$$\frac{1}{r!} \Phi^{(r)}_{x\ldots x}(x, x; \Delta x = -\varepsilon_i, \ldots, \Delta x = -\varepsilon_i)$$
$$= \sum_\ell \gamma_{1,\ell} \varepsilon_i^{\gamma_{2,\ell}} \varepsilon_i \ldots \varepsilon_i^{\gamma_{r+1,\ell}} = B(\varepsilon_i),$$

and (21) becomes

(23) $$A_i \tilde{\varepsilon}_{i+1} = \tilde{B}(\varepsilon_i) + \tilde{R}(\varepsilon_i),$$
hence
(24) $$\tilde{\varepsilon}_{i+1} = A_i^{-1}(\tilde{B}(\varepsilon_i) + \tilde{R}(\varepsilon_i)).$$

Let N denote an arbitrary Hölder norm. As $R(\varepsilon_i)$ is of degree $r+1$ in ε_i, we will be able to write

$$N(\varepsilon_{i+1}) = N(\tilde{\varepsilon}_{i+1})$$
(25)
$$\leq const.\ N(A^{-1}) \left[\sum_\ell N(\gamma_{1,\ell}) \ldots N(\gamma_{r+1,\ell}) + N(\rho_i) \right] (N(\varepsilon_i))^r,$$

where $N(\rho_i)$ tends to zero at the same time as ε_i.

The order of convergence is r, and convergence is always achieved if one starts with a value close enough to the solution. Our reasoning would be the same for iterations (5) and (6), and the same as above would be noted if we encountered a singular matrix A_i.

3.2 Singularity of the linear operator

In the sequel we assume matrix A to be singular. The matrices A_i may be regular or singular.

In order that equation (8') be compatible in $\tilde{x}$, we must have

(26) $$AA^{-1}\tilde{\varphi} = \tilde{\varphi},$$

where A^{-1} is the pseudo-inverse of A (see A. Korganoff and M. Pavel-Parvu [6]). This is an essential condition. We will assume it valid. Then (8') has an infinite set of solutions, and if iteration (4) converges, its limit point will be one of them, namely

(27) $$\tilde{x} = A^{-1}\tilde{\varphi} + (I - A^{-1}A)\tilde{\xi},$$

where $\tilde{\xi}$ is some vector parameter. For $\tilde{\xi}=0$ one obtains convergence to the best solution $A^{-1}\tilde{\varphi}$, which corresponds to the best solution of $\Phi(x,y) = 0$ in the sense that $||\Phi(x,y)|| = 0$ and $||y|| = min$ (y assuming the value x), where $||.||$ is the quadratic norm. x is then still a solution of $\Phi(x,x) = 0$, and hence a fixed point of iteration (4).

Let us consider iteration (5). It may be written in the form

(28) $$A_i \tilde{x}_{i+1} = A_i \tilde{x}_i + \tilde{\psi}_i,$$

and in the limit,

$$(29) \quad A \tilde{\tilde{x}} = A \tilde{x} + \tilde{\psi}.$$

In order that $\tilde{\tilde{x}}$ be a fixed point of (28), it is necessary that $\tilde{\psi}(x) = 0$. If our iteration process is convergent, its limit is a solution of $A \tilde{\tilde{x}} = A \tilde{x}$ (more precisely of $A\tilde{y} = A\tilde{x}$, $\tilde{y}$ assuming the value $\tilde{\tilde{x}}$), namely

$$(30) \quad \tilde{\tilde{x}} = A^{-1} A \tilde{x} + (I - A^{-1} A)\tilde{\xi}.$$

It is also one of the values

$$(31) \quad \tilde{\tilde{x}} = (I - A^{-1} A)\tilde{\xi} + (A^{-1} A)\tilde{\chi}.$$

It follows that $\tilde{\tilde{x}}$ is of an arbitrary form, and finally it is only compelled to satisfy $\tilde{\psi}(x) = 0$. This was obvious; however, relation (30) shows that by fixing $\tilde{\xi}$, one can fix the form of $\tilde{\tilde{x}}$. If for instance $\tilde{\xi} = 0$ (or $\xi = A^{-1} A\tilde{\xi}$), $\tilde{\tilde{x}}$ will be in the column space of A^{-1}, hence of A^*. This being so, x is a solution of $\psi(x) = 0$.

Let us consider iteration (6). It can be written in the form

$$(32) \quad A_i (\tilde{x}_{i+1} - \tilde{x}_i) = \tilde{\psi}_i,$$

and in the limit

$$(33) \quad A (\tilde{\tilde{x}} - \tilde{x}) = \tilde{\psi}.$$

As above, we must have $\tilde{\psi}(x) = 0$, and if our iteration process converges at all, its limit is a solution of $A(\tilde{\tilde{x}} - \tilde{x}) = 0$ (more precisely of $A(\tilde{y}-\tilde{x}) = 0$, $\tilde{y}$ assuming the value $\tilde{\tilde{x}}$), namely

$$(34) \quad \tilde{\tilde{x}} = \tilde{x} + (I - A^{-1} A)\tilde{\xi},$$

which is satisfied only if $\tilde{\xi}$ is null or of the form $A^{-1}A\xi$. This shows that, if the matrices A_i are singular, one must necessarily take the best solution of (32), namely

$$(35) \qquad \tilde{x}_{i+1} = \tilde{x}_i + A_i^{-1} \tilde{\psi}_i .$$

x will thus be a solution of $\psi(x) = 0$.

Let us specify that the singularity of the linear operators does not necessarily involve the singularity of x. Let us consider for instance the iteration

$$x_i\, x_{i+1} + x_{i+1}\, x_i = x_i^2 + a,$$

leading to a solution of $x^2 - a = 0$, i.e. to a square root of a. The matrix A is then expressed as

$$A = x \otimes I + I \otimes x^T ,$$

and its eigenvalues are $\lambda_i + \lambda_j$, where λ_i and λ_j are two arbitrary eigenvalues of x. A is obviously singular if some λ_i is null, hence if x is singular, but also if simply two of the eigenvalues of x have opposite signs.

The study of convergence still remains. If A_i approximates singularity, obviously x_{i+1} increases indefinitely in norm. The pseudo-inverse of A is defined only for exact singularity, thus for x, and there is no continuity in passing from the inverse A_i^{-1} to the pseudo-inverse A^{-1} (no more, incidentally, than in passing from

one pseudo-inverse to another when a complementary eigenvalue
vanishes), or from x_{i+1} to x_i, i.e. from $\tilde{x}_{i+1}$ to $\tilde{x}_i$. This drawback
may be palliated by creating or widening the domain around x where
the matrices $A(x)$ are singular. It is enough in a direct method of
solution for a linear system, e.g. for $A_i\,\tilde{x}_{i+1} = \tilde{\varphi}_i$, to increase
the value of the numerical zero for a linear independence test.
Things seem more delicate with an iterative method. This is because
it is difficult to test the singularity other than by the unmeasur-
ed growth of $\tilde{x}_{i+1}$. Hence one must try to separate the part due to
small eigenvalues from that due to big (or relatively big) ones in
order to eliminate their influence.

To simplify notations let us omit the subscripts i, and
consider an iterative method which proceeds by a splitting
$A = M - N$, and let us assume A to be singular (M being regular).
Let v_1 and v_2 be the matrices of eigenvectors of $I - M^{-1} A$, and I
and λ_2 the diagonal matrices of the corresponding eigenvalues:

$$(I - M^{-1} A)\, v_1 = v_1 ,$$
$$(I - M^{-1} A)\, v_2 = v_2\, \lambda_2 ;$$

thus

$$v_1 = (I - A^{-1} A)\, \bar{\omega} ,$$
$$v_2 (I - \lambda_2) = M^{-1} A\, v_2 ,$$

$\bar{\omega}$ being an arbitrary matrix. Let $\tilde{x}^{(0)}$ be an initial value; $\tilde{x}^{(0)}$ and
$M^{-1}\tilde{\varphi}$ may be expressed as

$$\tilde{x}^{(0)} = v_1 \tilde{\sigma}_1 + v_2 \tilde{\sigma}_2,$$

$$M^{-1}\tilde{\varphi} = v_1 \tilde{\theta}_1 + v_2 \tilde{\theta}_2,$$

where $\tilde{\sigma}_1$, $\tilde{\sigma}_2$, $\tilde{\theta}_1$ and $\tilde{\theta}_2$ are column vectors. One is then able to prove the following: if the equation $A\tilde{x} = \tilde{\varphi}$ is not compatible, and if the moduli of the elements of λ_2 are smaller than one (see Appendix 2), then

$$\lim_{j \to \infty} \tilde{x}^{(j+1)} = v_1 \tilde{\sigma}_1 + \lim_{j \to \infty} (j+1)\, v_1 \tilde{\theta}_1 + v_2 (I-\lambda_2)^{-1}\tilde{\theta}_2,$$

with the result that from some value of j onward

$$\tilde{x}^{(j+1)} - \tilde{x}^{(j)} \sim v_1 \tilde{\theta}_1,$$

and

$$\tilde{z}^{(j+1)} = (j+1)\,\tilde{x}^{(j)} - j\,\tilde{x}^{(j+1)} \sim (I - A^{-1}A)\,\bar{\omega}\tilde{\sigma}_1 + v_2 (I - \lambda_2)^{-1}\tilde{\theta}_2.$$

It follows that $\tilde{z} = \lim_{j \to \infty} \tilde{z}^{(j+1)}$ is a solution of

$$A\tilde{z} = \tilde{\varphi} - M(I - A^{-1})\,\bar{\omega}\tilde{\theta}_1.$$

If equation $A\tilde{x} = \tilde{\varphi}$ were compatible, then $\tilde{x} = \lim_{j \to \infty} \tilde{x}^{(j+1)}$ (see H.B. Keller [7]), and $\tilde{z}$ will be a solution of the equation. In order to obtain the best solution $A^{-1}\tilde{\varphi}$ using an adequate iterative method (see Appendix 2), we would start with an initial value, say $\tilde{x}^{(0)}$, located in the column space of A^*. This means $\tilde{x}^{(0)}$ is of the form $A^*\tilde{\xi}$, e.g. $\tilde{x}^{(0)} = 0$.

We can then say that, if A tends to singularity but is regular (the singularity may take place at the ith step), and if (at this i^{th} step) $A\tilde{x} = \tilde{\varphi}$ is incompatible, the norm of $\tilde{x}^{(j+1)}$, but not the one of $\tilde{z}^{(j+1)}$ strongly increases; $\tilde{z}^{(j+1)}$ veers away from being the solution of $A\tilde{x} = \tilde{\varphi}$ to being a solution of $A\tilde{x} = \tilde{\varphi} - M (I-A^{-1} A) \tilde{\omega\theta}_1$. This means that it is desirable - although expensive - to make the equations compatible, i.e. to work with

(36)
$$A_i^* A_i \tilde{x}_{i+1} = A_i^* \tilde{\varphi}_i,$$
$$A_i^* A_i \tilde{x}_{i+1} = A_i^* A_i \tilde{x}_i + A_i^* \tilde{\psi}_i,$$
$$A_i^* A_i (\tilde{x}_{i+1} - \tilde{x}_i) = A_i^* \tilde{\psi}_i,$$

(or with any other compatible equations), and to test for convergence of $\tilde{z}^{(j+1)}$, in order to avoid irregularities of the variations of x, due to weak incompatibilities originating from rounding-off errors. This is a point worthwhile studying thoroughly and testing numerically. It would be equally worthwhile studying how to proceed with projection or gradient iterative methods (see for such methods A. Korganoff and M. Pavel-Parvu [6]).

Summarizing, one can have numerical convergence only if one places oneself permanently inside a domain where the considered matrices are numerically singular and have the same rank, which means placing oneself directly in the subspace where the pseudo-inverse is defined. Continuity of the pseudo-inverse will then be assured. This point also would deserve more careful theoretical study.

We will hence start with equations (36). They are of the same form as equations (8), (28) and (32), if one replaces A_i by $A_i^* A_i$, $\tilde{\varphi}_i$ by $A_i^{*}\tilde{\psi}_i$ and $\tilde{\psi}_i$ by $A_i^{*}\tilde{\psi}_i$. One associates with equations (36) matricial iteration functions analogous to (4), (5) and (6). Taking the relations

$$
(37) \quad \begin{aligned} A_i^* A_i &= (\sum_k \alpha_k^* \circledast (\beta_k^T)^*)(\sum_\ell \alpha_\ell \circledast \beta_\ell^T) \\ &= \sum_{k,\ell} (\alpha_k \alpha_\ell) \circledast (\beta_\ell \beta_k^*)^T \end{aligned}
$$

into account, one gets for instance for the first equation

$$(38) \quad \sum_{k,\ell} \alpha_k^*(x_i)\, \alpha_\ell(x_i)\, x_{i+1}\, \beta_\ell(x_i)\, \beta_k^*(x_i) = \sum_k \alpha_k^*(x_i)\varphi(x_i)\, \beta_k^*(x_i).$$

This means, as above, replacing in (4) α_k by $\alpha_k^* \alpha_\ell$, β_k by $\beta_\ell \beta_k^*$ and φ by $\sum_k \alpha_k^* \varphi \beta_k^*$. To simplify notations we will return to (8), (28), and (32), where the equations will be compatible and A hermitian.

This being admitted, convergence has to be studied on the iterations

$$(39) \quad \tilde{x}_{i+1} = A_i^{-1}\tilde{\varphi}_i + (I - A_i^{-1} A_i)\tilde{\xi}_i,$$

$$(40) \quad \tilde{x}_{i+1} = A_i^{-1} A_i \tilde{x}_i + A_i^{-1}\tilde{\psi}_i + (I - A_i^{-1} A_i)\tilde{\xi}_i,$$

$$(41) \quad \tilde{x}_{i+1} = \tilde{x}_i + A_i^{-1}\tilde{\psi}_i,$$

(and not on (4), (5) and (6)). This is so because (16) for instance gives

$$\tilde{\varepsilon}_{i+1} = - A_i^{-1} B_i \varepsilon_i + (I - A_i^{-1} A_i) \tilde{\zeta}_i$$

and $\tilde{\zeta}_i$, not necessarily equal to 0 or to $\tilde{\xi}_i$, is thus undetermined. Let us revert to our iteration: one can easily see that if $\tilde{\xi}_i \neq 0$, (39) and (40) lead to linear iterations only, except when the derivatives of $A^{-1}A$ are zero. Besides, convergence will be achieved only if the $\tilde{\xi}_i$ are independent of i. Let us specify that, except for the value 0, $\tilde{\xi}$ is not easily controlled. In cases of linear convergence iteration (41), but not (40), seems to be divergent (at first sight and errors excepted). As it is difficult to exhibit necessary and sufficient conditions of convergence (all matrices involved cannot be put, with certain exceptions, into tensor product form), we will restrict our research to sufficient conditions merely. As a specific example we will study iteration (39).

a) Linear convergence

Replacing in (39) x_i by $x - \varepsilon_i$:

(42) $\quad \tilde{x} - \tilde{\varepsilon}_{i+1} = \left[A(x-\varepsilon_i) \right]^{-1} \tilde{\varphi}(x-\varepsilon_i) + \{ I - \left[A(x-\varepsilon_i) \right]^{-1} A(x-\varepsilon_i) \} \tilde{\xi}_i$.

$(A(x))^{-1}$ is a continuous function of x (at least that is what we have assumed), so that we can write

(43) $\quad \left[A(x-\varepsilon_i) \right]^{-1} = A^{-1}(x-\varepsilon_i) = A^{-1}(x) + (A^{-1})'(x;-\varepsilon_i) + \eta \lVert \varepsilon_i \rVert$,

$\lVert \eta \rVert$ tending to zero with $\lVert \varepsilon_i \rVert$ (see Appendix 1). As likewise $(\tilde{\varphi})' = (\tilde{\varphi'})$, we get, taking $\tilde{\xi}_i = \tilde{\xi}$:

$$\tilde{\varepsilon}_{i+1} \sim - A^{-1}(x)\, \tilde{\varphi}'(x;-\varepsilon_i) - (A^{-1})'(x;-\varepsilon_i)\, \tilde{\varphi}(x)$$
(44)
$$+\left[A^{-1}(x)\, A'(x;-\varepsilon_i) + (A^{-1})'(x;-\varepsilon_i)\, A(x)\right] \tilde{\xi}.$$

Let us consider each derivative separately, and denote by N a Hölder norm — for which, besides,

(45) $$N(\alpha \circledast \beta) = N(\alpha)\, N(\beta).$$

Taking (7) into account, we then have

(46) $$A'(x;-\varepsilon_i) = \sum_k \alpha'_k(x;-\varepsilon_i) \circledast \beta_k^T(x) + \sum_k \alpha_k(x) \circledast \beta_k'^{T}(x;-\varepsilon_i),$$

so that $A'(x;-\varepsilon_i)$ is of the form

(47)
$$A'(x;-\varepsilon_i) = \sum_k \sum_\ell \gamma_{k,1}^{(\ell)}(x)\, \varepsilon_i\, \gamma_{k,2}^{(\ell)}(x) \circledast \beta_k^T(x)$$
$$+ \sum_k \sum_m \alpha_k(x) \circledast (\delta_{k,2}^{(m)})^T(x)\, \varepsilon_i^T\, (\delta_{k,1}^{(m)})^T(x)$$

and, simplifying notation:

(48)
$$N(A') \leq \{\sum_k [\text{const.}\, N(\beta_k) \sum_\ell N(\gamma_{k,1}^{(\ell)})\, N(\gamma_{k,2}^{(\ell)})$$
$$+ \text{const.}\, N(\alpha_k) \sum_m N(\delta_{k,1}^{(m)})\, N(\delta_{k,2}^{(m)})]\}\, N(\varepsilon_i).$$

As for $\varphi'(x;-\varepsilon_i)$, it is of the form

(49) $$\varphi'(x;-\varepsilon_i) = \sum_n \mu_{n,1}(x)\, \varepsilon_i \mu_{n,2}(x),$$

and

(50) $\quad N(\tilde{\varphi}') = N(\varphi') \leq const. \left[\sum_n N(\mu_{n,1}) N(\mu_{m,2})\right] N(\varepsilon_i).$

Remains $(A^{-1})'(x;-\varepsilon_i)$; its expression in terms of A' (see Appendix 1) is:

(51)
$$(A^{-1})' = - A^{-1} A' A^{-1} + A^{-1} (a^*)^{-1} (a^*)' (I-AA^{-1})$$
$$+ (I - A^{-1} A) (b^*)' (b^*)^{-1} A^{-1},$$

where a and b are bases of the column and the row spaces, respectively, of A, and are in particular its independent columns and its independent rows, respectively. Hence

(52) $\quad a = AP_1 \begin{bmatrix} I \\ 0 \end{bmatrix} = AQ_1, \quad b = \begin{bmatrix} I & 0 \end{bmatrix} P_2 A = Q_2 A,$

where P_1 and P_2 are permutation matrices, and

(53) $\quad a' = A'Q_1, \quad b' = Q_2 A'.$

It then follows that

(54) $\quad N((A^{-1})') \leq [. \ . \ .] N(\varepsilon_i),$

and finally

(55) $\quad N(\varepsilon_{i+1}) = N(\tilde{\varepsilon}_{i+1}) \leq [. \ . \ .] N(\varepsilon_i).$

One deduces from this a sufficient condition of convergence.

b) **Superlinear convergence**

If one reverts to (44), and assumes $\tilde{\xi} = 0$, it can be seen that the conditions for achieving quadratic convergence ar no longer identical to those for non-singular A. We must have, as a matter of fact,

(56) $\qquad -A^{-1}(x)\,\tilde{\varphi}'(x;\Delta x) - (A^{-1})'(x;\Delta x)\,\tilde{\varphi}(x) \equiv 0,$

i.e. on account of (51), (52) and (53)

$$- A^{-1}\tilde{\varphi}' + A^{-1} A' A^{-1} \tilde{\varphi} - A^{-1}(Q_1^* A^*)^{-1} Q_1^*(A^*)'(I - AA^{-1})\tilde{\varphi}$$

$$- (I - A^{-1} A)(A^*)' Q_2^*(A^* Q_2^*)^{-1} A^{-1}\tilde{\varphi} \equiv 0.$$

Hence, on account of (26) and (27),

(57) $\qquad - A^{-1}\tilde{\varphi}' + A^{-1} A' \tilde{x} - (I - A^{-1} A)(A^*)' Q_2^*(A^* Q_2^*)^{-1}\tilde{x} \equiv 0.$

If A is not singular, this condition leads back to the first condition (20), since the only remaining equalities are then $A'\tilde{x} - \tilde{\varphi}' = \varphi'$, $\tilde{\varphi}'(x,x;\Delta x) = B\Delta\tilde{x} \equiv 0$, i.e. $B \equiv 0$, and $\varphi'(x,x;\Delta x) \equiv 0$. Notice that identity (57), meant to be an equation in $A'\tilde{x} - \tilde{\varphi}'$, is incompatible. Thus the quadratic convergence condition cannot be written $\varphi'(x,x;\Delta x) - F(x,\Delta x) \equiv 0$, but instead $A^{-1}B\Delta\tilde{x} - G(x,\Delta x) \equiv 0$. As G is generally not (with rare exceptions only) of the form $D\Delta\tilde{x}$, one is not able to obtain a formulation of linear convergence analogous to (16) or (17). This explains a posteriori the difficulty encountered in obtaining necessary and sufficient conditions for linear convergence when A is singular.

Under these circumstances convergence of order r of the iterations (39), (40) and (41) would be studied in the same way as in 3.1.b and 3.2.a. Convergence is always achieved for these three iterations on condition that one starts with a value close enough to the solution. The difficulty seems here to consist in determining iterations where the order of convergence is $r > 1$.

3.3 Nullity of the linear operator at the solution

This case is analogous to the scalar one, where for instance in Newton's method $f'(x) = 0$; f/f' is then undetermined. One has under these circumstances multiple roots. We will return to these in the next section.

For our present purposes, continuity of A^{-1} is no longer given. $A^{-1}(x_i)$ tends to infinity and is equal to the null operator at the solution only.

Let us study iteration (4) again or more precisely (8), i.e.

(58) $$A_i \tilde{x}_{i+1} = \tilde{\varphi}_i ,$$

and let us expand A_i and $\tilde{\varphi}_i$. One obtains

(59) $$\left[A(x) + A'(x; -\varepsilon_i) + \frac{1}{2!} A''(x; -\varepsilon_i, -\varepsilon_i) + \ldots \right] (\tilde{x} - \tilde{\varepsilon}_{i+1})$$
$$= \tilde{\varphi}(x) + \tilde{\varphi}'(x; -\varepsilon_i) + \frac{1}{2!} \tilde{\varphi}''(x; -\varepsilon_i, -\varepsilon_i) + \ldots$$

As $A(x) \equiv 0$, convergence requires

(60) $$\tilde{\varphi}(x) \equiv 0, \quad A'(x; \Delta x) \tilde{x} - \tilde{\varphi}'(x; \Delta x) \equiv 0.$$

At A^{-1} one finds the same undetermination as for f/f' in the scalar case. Let us assume, for simplicity's sake, that A_i is regular for every i. (If it were singular, one would proceed as in section 3.2 and consider the pseudo-inverse of A_i). Then $\tilde{x}_{i+1} = A_i^{-1} \tilde{\varphi}_i$, and $\tilde{x}_{i+1}$ tends (with certain exceptions) to a finite value. Under these circumstances, the solution of (58) is

$$\tilde{x} = \lim_{i \to \infty} \tilde{x}_{i+1} , \tag{61}$$

and not $\tilde{x} = A^{-1} \tilde{\varphi}$ which is null.

Let us revert to (59). One has, with a third order approximation:

$$A'(x;-\varepsilon_i)\tilde{\varepsilon}_{i+1} = \frac{1}{2!} A''(x;-\varepsilon_i,-\varepsilon_i) \tilde{x} - \frac{1}{2!} \tilde{\varphi}''(x;-\varepsilon_i,-\varepsilon_i). \tag{62}$$

Let us assume that $A'(x;\Delta x)$ as well as the right-hand side of (62) are not identically null, and that $A'(x;-\varepsilon_i)$ is regular for every i (except, obviously at the limit, i.e. for $\varepsilon_i=0$). Equality (62) indicates linear convergence. This is what we will try to prove.

After simplification of notation one first infers

$$\tilde{\varepsilon}_{i+1} = \frac{1}{2!} (A')^{-1} (A'' \tilde{x} - \tilde{\varphi}''), \tag{63}$$

and then, denoting by N a Hölder norm,

$$N(\varepsilon_{i+1}) = N(\tilde{\varepsilon}_{i+1}) \leq const. \, N((A')^{-1}) \left[const. \, N(A'') \, N(\tilde{x}) + N(\tilde{\varphi}'') \right]. \tag{64}$$

There exist two constants such that

(65) $\quad const.\ N((A')^{-1}) \leq ||(A')^{-1}||_2 \leq const.\ N((A')^{-1}),$

where $||.||_2$ is the spectral norm:

(66)
$$||(A')^{-1}||_2 = [\lambda(((A')^{-1})^* (A')^{-1})_{max}]^{1/2}$$
$$= [\lambda(A'\ (A')^*)_{min}]^{-1/2}$$

(λ denoting eigenvalues).

The eigenvalues of a matrix are continuous functions of its elements. Actually $\lambda(A'(A')^*)$ is, taking (47) into account, a continuous function of the elements of ε_i. If ε_i is small, one can write in a first approximation

(67) $\left[\lambda(A'(A')^*)_{min}\right] = \sum_{j,k,l,m} h_{j,k,l,m} (\varepsilon_i)_{j,k} (\varepsilon_i)_{l,m},$

where $h_{j,k,l,m}$ is a scalar function of x. The matrix $A'(A')^*$ being hermitian positive definite, its eigenvalues are real and positive, and

(68) $\quad\quad\quad\quad \lambda(A'(A')^*)_{min} = \tilde{\varepsilon}_i^* H \tilde{\varepsilon}_i,$

where H, too, is a positive definite hermitian matrix. In this case

(69) $\quad\quad\quad\quad m_H \leq \dfrac{\tilde{\varepsilon}_i^* H \tilde{\varepsilon}_i}{\tilde{\varepsilon}_i^* \tilde{\varepsilon}_i} \leq M_H,$

where m_H and M_H are the bounds of the spectrum of H. One infers:

(70) $\quad N((A')^{-1}) \leq const.||(A')^{-1}||_2 \leq const. \dfrac{1}{\sqrt{m_H} \, \tilde{\varepsilon}_i^* \, \tilde{\varepsilon}_i} = \dfrac{const.}{\sqrt{m_H}} \dfrac{1}{N(\varepsilon_i)}.$

As for $N(A'')$ and $N(\tilde{\varphi}'')$, one could easily prove, as in (48) and (50), that they are of the form $[\ldots](N(\varepsilon_i))^2$. One finally has

(71) $\qquad\qquad\qquad N(\varepsilon_{i+1}) \leq [\ldots] \, N(\varepsilon_i) ,$

and convergence is really linear.

Linear convergence could also be achieved with a term of degree t in ε_i on the left-hand side of (62), i.e. $A^{(t)}(x;-\varepsilon_i,\ldots,-\varepsilon_i)$ (in which case one would have $A' \equiv 0,\ldots, A^{(t-1)} \equiv 0$), and with terms of degree $t+1$ in ε_i on the right-hand side. It would be worthwhile generalizing the previous arguments – with difficulties, however, arising in obtaining a lower bound for $N((A^{(t)})^{-1})$ – and, next, extending them to iterations of order $r > 1$.

3.4 Multiple roots

Let $f(x) = 0$ be a matricial polynomial equation of degree ν, and let $x_{(0)}$ be a root of multiplicity μ. By definition:

(72) $\quad f'(x_{(0)};\Delta x) \equiv 0,\ldots,f^{(\mu-1)}(x_{(0)};\Delta x,\ldots,\Delta x) \equiv 0,$

$\qquad f^{(\mu)}(x_{(0)};\Delta x,\ldots,\Delta x) \neq 0 .$

Let us then consider iteration (5), which is better suited to this problem than iteration (4). In the scalar case, $\psi(x)$ is of the form $\psi(x) = h(x) f(x)$, where $h(x)$ is a function which does not vanish for $x_{(0)}$. In the matricial case $f(x)$ cannot be necessarily factorized, and $\psi(x)$ will simply be a function of f vanishing therewith:

$$\psi(x) = \psi(f(x)),$$
(73)
$$\psi(0) = 0.$$

If one expands $\psi(x)$ around $x_{(0)}$, and takes (72) into account, one may write:

(74)
$$\psi(x) = \frac{1}{\mu!} \psi_x^{(\mu)}(x_{(0)}; x - x_{(0)}, \ldots, x - x_{(0)}) + \eta(x)$$
$$= \frac{1}{\mu!} \psi_f'(x_{(0)}; f^{(\mu)}(x_{(0)}; x - x_{(0)}, \ldots, x - x_{(0)})) + \eta(x),$$

$\eta(x)$ tending to zero like $||x - x_{(0)}||^{\mu+1}$.

Replacing in (5) $\psi(x)$ by (74), and setting as usual $\varepsilon = x - x_{(0)}$, we will have with first-order approximation:

(75)
$$\sum_k \alpha_k(x_{(0)}) \varepsilon_{i+1} \beta_k(x_{(0)}) = \sum_k \alpha_k(x_{(0)}) \varepsilon_i \beta_k(x_{(0)})$$
$$+ \frac{1}{\mu!} \psi_f'(x_{(0)}; f^{(\mu)}(x_{(0)}; -\varepsilon_i, \ldots, -\varepsilon_i)).$$

ψ_f' being homogeneous in $f^{(\mu)}$, the latter being itself homogeneous and of degree μ in ε_i, the iteration appears to be linear and divergent. For convergence it is necessary that $A(x) = \sum_k \alpha_k(x) \otimes \beta_k^T(x)$ and that its derivatives up to order $\mu - 2$ vanish at $x_{(0)}$:

252   Iteration Functions for Solving Polynomial Matrix Equations

one is referred to section 3.3. What actually occurs is: like in Newton's method, when

(76) $$f'(x_i; x_{i+1}) = f'(x_i; x_i) - f(x_i),$$

convergence, which is quadratic for an isolated root, becomes linear for a multiple one.

The derivatives of A vanishing up to order $t \geq \mu-2$, the right-hand side of iteration (5) (and not A) should have derivatives equal to zero up to order $t + r$, in order to have convergence of order r.

3.5 Commutativity

Commutativity is interesting as it simplifies calculus considerably.

Let us take again iteration (4)

$$\sum_k \alpha_k(x_i) \, x_{i+1} \, \beta_k(x_i) = \varphi(x_i),$$

and let us assume that x_{i+1}, for instance, commutes with all the $\beta_k(x_i)$ (in which case, besides, the β_k and x are square matrices). Iteration (4) will be written

(77) $$\left[\sum_k \alpha_k(x_i) \, \beta_k(x_i)\right] x_{i+1} = \varphi(x_i).$$

Let

(78) $$\sum_k \alpha_k(x_i) \, \beta_k(x_i) = a(x_i) = a_i.$$

If a is square and regular for every i (or if a_i is an (m,n)-matrix, $m > n$, of maximal rank, such that (77) be compatible, hence such that $a_i^{-1} a_i \varphi_i = \varphi_i, \forall i$)

(79) $$x_{i+1} = a_i^{-1} \varphi_i ,$$

and, in order that commutativity be effective, one should have

(80) $$a_i^{-1} \varphi_i . (\beta_k)_i = (\beta_k)_i . a_i^{-1} \varphi_i, \forall k .$$

In addition it is necessary that (80) be satisfied for $i+1$, if it is for i. This implies

(80') $$a^{-1}(a_i^{-1}\varphi_i) (a_i^{-1}\varphi_i) \beta_k (a_i^{-1} \varphi_i)$$
$$= \beta_k (a_i^{-1} \varphi_i) a^{-1} (a_i^{-1} \varphi_i) (a_i^{-1}\varphi_i),$$

and even more: (80) must be satisfied in the limit, i.e. at the solution (and if a is rectangular, (77) must obviously be compatible for the solution).

Let us consider as an example the iteration

(81) $$x_i x_{i+1} + x_{i+1} x_i = x_i^2 + a,$$

converging to a square root of a. The commutativity relation (80) can be written

(82) $$x_i^{-1} (x_i^2 + a) x_i = x_i x_i^{-1} (x_i^2 + a),$$

hence, if x_i is regular,

(83) $$a\, x_i = x_i\, a .$$

Under this condition,

(84) $$x_{i+1} = \frac{1}{2} x_i^{-1} (x_i^2 + a) ,$$

and relation (80') - reduced to (84) - now reads

$$a\, x_i^{-1}(x_i^2 + a) = x_i^{-1} (x_i^2 + a)\, a,$$

which, taking (84) into account, is always satisfied. It is then sufficient that x_0 commutes with a in order that (83) holds for any i. This is obvious from (83). In order that $\lim x_{i+1}$ be a solution, it is necessary that the solution x, too, commutes with a. This, however, follows from $x^2 = a$, which yields $x = x^{-1}a = ax^{-1}$. One then verifies that:

$$x^2 = \frac{1}{4} x^{-1} (x^2+a)\, x^{-1} (x^2+a) = \frac{1}{4} (x^2+a+x^{-1}ax+x^{-1}ax^{-1}a) = a.$$

Let us now suppose that $a(x)$ is square singular (or rectangular and of non-maximum rank, or rectangular and of maximum rank $n(m<n)$). Notice that the singularities of a_i and A_i do not necessarily coincide. Thus, as concerns (81), a_i is singular with x_i, when A_i is singular with $x_i \otimes I + I \otimes x_i^T$ (see section 3.2). Let us assume that the compatibility conditions $aa^{-1}\varphi = \varphi$, $a_i a_i^{-1}\varphi_i = \varphi_i$ are satisfied((77) being made compatible, if necessary). Commutativity conditions (80) and (80') subsist if one restricts oneself

to the best solution of (77), but cannot always be simplified as in (83). If one considers (82) again, one has, taking the compatibility condition $x_i \, x_i^{-1} \, a = a$ into account,

(85) $$x_i^{-1}(x_i^2 + a) \, x_i = x_i^2 + a \; .$$

Relation (80'), which may be written

(86) $$\left[x_i^{-1}(x_i^2 + a)\right]^{-1} \left[x_i^{-1}(x_i^2 + a) \, x_i^{-1}(x_i^2 + a) + 4a\right] x_i^{-1}(x_i^2 + a)$$
$$= x_i^{-1}(x_i^2 + a) \, x_i^{-1}(x_i^2 + a) + 4a$$

can no longer be simplified so easily, and is no longer systematically verified.

One realizes the problems thus arising. To these one must add those connected with the loss of commutativity during the iterations (see P. Laasonen [8]). One should finally adapt the convergence studies previously done.

4. Convergence. General iteration functions

We are now concerned with iterations of the form (3). For simplicity we suppose that we deal with a single function $g(x)$, such that:

(87) $$\sum_{k_1} \alpha_{1,k_1}(x_i, g(x_i)) \, x_{i+1} \, \beta_{1,k_1}(x_i, g(x_i)) = \varphi_1(x_i, g(x_i)),$$
$$\sum_{k_2} \alpha_{2,k_2}(x_i) \, g(x_i) \, \beta_{2;k_2}(x_i) = \varphi_2(x_i) \; .$$

Convergence is studied on the first relation (87), thus in the same way as for ordinary iterations. Let us mention however, that the functions are composite ones, and that their derivatives have a particular form. For the first derivative of an α, for instance, one has

$$\alpha_x'(x,g;\Delta x) + \alpha_g'(x,g;\Delta g = g_x'(x;\Delta x)).$$

The total derivative is obviously linear and homogeneous in Δx.

The second relation (87) still plays a non-negligible part. On the one hand, if the corresponding linear operator, which will be denoted by A_2, is singular, several determinations of $g(x)$ — namely

(88) $$\tilde{g} = A_2^{-1} \tilde{\varphi}_2 + (I - A_2^{-1} A_2) \tilde{\xi}_2$$

are possible (compatibility being assured). On the other hand, the derivative of $\tilde{g}$, expressed by (see (44))

(89) $$\tilde{g}' = (A_2^{-1})' \tilde{\varphi}_2 + A_2^{-1} \tilde{\varphi}_2' - (A_2^{-1} A_2)' \tilde{\xi}_2,$$

may, depending on the value of $\tilde{\xi}_2$, influence the order of convergence of the iteration.

This being settled, the study of convergence may proceed along the same lines as in section 3.

5. Iteration Functions

One can generalize many scalar iteration functions. Actually all — or, to be careful, almost all — those which are directly or indirectly based upon a series expansion around the solution of the inverse function $x(f)$ of the left-hand side of the equation $f(x) = 0$ to be solved, and which call for f and its derivatives. In short all those which are based upon extrapolation formulas in the sense in which Taylor's formula is one. On the contrary, iterations based upon interpolation formulas, for which commutativity plays an important role (we assume here not to be in a case of commutativity of matricial iterations) do not admit generalizations, at least on first approach. This section will hence be modeled, as far as needed, on the scalar case, for which we refer the reader to J.F. Traub [5].

The formulation of matricial iteration functions is not identical to that of scalar iteration functions, but the transcription technique is easy. The order of convergence is maintained if matrix A is non-singular. If A is singular at the solution, the order of convergence becomes linear(with certain exceptions) and the iteration may then be convergent or divergent, unless the contrary is proved.

Let us consider an iteration function with quadratic convergence when A is regular. One will have (see (20) and (14))

(90) $$\phi'_x(x,x;\Delta x) \equiv 0,$$

or also

(90') $$A'\tilde{x} - \tilde{\phi}' \equiv 0.$$

Taking (57) into account (let us recollect that $\xi = 0$) it will result that the iteration remains quadratically convergent for singular A if

(91) $$(I - A^{-1}A)(A^*)' Q_2^* (A^* Q_2^*)^{-1} \tilde{x} \equiv 0,$$

hence, in particular, ($x = 0$ being excluded) if the columns of $(A^*)'$ lie in the column space of A^*:

(92) $$(A^*)' = A^* S(x, \Delta x).$$

But this is not at all obligatory[*)].

Let us examine identity (91) in more detail. To this end let

(93) $$Q_2^* (A^* Q_2^*)^{-1} \tilde{x} = \tilde{z}.$$

Taking (47) into account, one obtains after some simplification of notation:

$$(I - A^{-1}A)(A^*)' \tilde{z} = -(I - A^{-1}A)(\Sigma\Sigma\gamma_2^* \Delta x^* \gamma_1^* \ominus \bar{\beta}$$

$$+ \Sigma\Sigma\alpha^* \ominus \bar{\delta}_1 \Delta \bar{x} \bar{\delta}_2) \tilde{z}$$

$$= -(I - A^{-1}A)(\Sigma\Sigma\gamma_2^* \Delta x^* \gamma_1^* z\beta^* + \Sigma\Sigma\alpha^* z\delta_2^* \Delta x^* \delta_1^*)$$

(94) $$= -(I - A^{-1}A)(\Sigma\Sigma\gamma_2^* \ominus \bar{\beta} z^T \bar{\gamma}_1 + \Sigma\Sigma\alpha^* z \delta_2^* \ominus \bar{\delta}_1) \widetilde{\Delta x}^*,$$

which will be zero for all Δx, if:

[*)] Let us specify (see Appendix 1) that
$$A' = a' a^{-1} A - Ab^{-1} \gamma' a^{-1} A + Ab^{-1} b'.$$

(94') $(I - A^{-1}A) \ (\Sigma\Sigma\gamma_2^* \otimes \bar{\beta} \ z^T \ \bar{\gamma}_1 + \Sigma\Sigma\alpha^* \ z \ \delta_2^* \otimes \bar{\delta}_1) \equiv 0,$

or else, if:

(95)
$$\Sigma\Sigma\gamma_2^* \otimes \bar{\beta} \ z^T \ \bar{\gamma}_1 + \Sigma\Sigma\alpha^* \ z \ \delta_2^* \otimes \bar{\delta}_1 = A^* T$$
$$= (\Sigma\alpha^* \otimes \bar{\beta}) \ T,$$

where T is an arbitrary matrix (of proper order). Relation (95) is nothing but the transcription of (92).

As an example let us consider (81), the convergence of which is quadratic if

(96) $\qquad A = x \otimes I + I \otimes x^T$

is regular, and let us assume that x has eigenvalue zero or two opposite eigenvalues. A is then singular. One has:

(97)
$$(A^*)' \ \tilde{z} = (\Delta x^* \otimes I + I \otimes \Delta\bar{x}) \ \tilde{z}$$
$$= (z \otimes I + I \otimes z^T) \ \Delta\tilde{x}^*.$$

Convergence will remain quadratic, if

(98) $\qquad z \otimes I + I \otimes z^T = (x^* \otimes I + I \otimes \bar{x}) \ T,$

where

(98') $\qquad \tilde{z} = Q_2^* \left[(x^* \otimes I + I \otimes \bar{x}) \ Q_2^* \right]^{-1} \tilde{x},$

and where the matrix T is undetermined. One sees that it does not follow that relation (98) has to be verified.

We will assume in what follows that A is regular.

5.1 General theorems

Two theorems are immediately transcribed. The first one refers to the composition of iterations. Let us denote by $\phi_1(x,y)$ the first ordinary iteration function ϕ (see (4')) of order r_1, and by $\phi_2(x,y_2)$ the second iteration function of order r_2. The iteration function

$$(99) \qquad \phi_3(x,y) = \phi_1(y_2(x),y)$$

is of order $r_1 r_2$. This is easily inferred from the convergence relation

$$N(\varepsilon_{i+1}) \leq [\ldots] \, (N(\varepsilon_i))^{r_1}$$

for the first iteration, and from

$$N(\varepsilon_i) \leq [\ldots] \, (N(\varepsilon_{i-1}))^{r_2}$$

for the second one. Extension to more than two iterations is obvious. The same is valid, subject to conditions indicated in section 4, for general iteration functions.

The second theorem refers to what J.F. Traub [5] calls "recursively formed iteration functions". Let $\phi_1(x,y_1)$ be an ordinary iteration function of order r. The iteration function

(100) $\quad \phi_2(x,y) = f'_x(x;y) - f'_x(x;y_1(x)) + f(y_1(x))$,

where $f(x)$ is a function for which a zero is wanted, is of order $r+1$.

Let us prove it for $r = 1$. One has:

$$(\phi_2)'_x(x,x;\Delta x) = f''_{xx}(x;x,\Delta x) - f''_{xx}(x;y_1,\Delta x)$$
(101)
$$- f'_x(x;y'_1(x;\Delta x)) + f'_x(y_1;y'_1(x;\Delta x)),$$

which is identically zero in a fixed point, since there $y_1(x) = x$.

Let us turn to $r = 2$. One has:

$$(\phi_2)''_{xx}(x,x;\Delta x, \Delta x) = f'''_{xxx}(x;x,\Delta x, \Delta x) - f'''_{xxx}(x;y_1,\Delta x, \Delta x)$$

$$- 2f'''_{xx}(x;y'_1(x;\Delta x), \Delta x)$$

$$- f'_x(x;y''_1(x;\Delta x, \Delta x))$$

$$+ f'''_{xx}(y_1;y'_1(x;\Delta x), y'_1(x;\Delta x))$$

(102) $$+ f'_x(y_1;y''_1(x;\Delta x, \Delta x)),$$

which is identically zero since $y_1 = x$, and, ϕ_1 being of order 2:

$$(\phi_1)'_x(x,x;\Delta x) + (\phi_1)'_{y_1}(x,x;\Delta y_1)$$
(103)
$$= (\phi_1)'_{y_1}(x,x;\Delta y_1) \equiv 0.$$

This means (see (12)),

(103') $A \tilde{y}'_1(x;\Delta x) \equiv 0,$

and, if A is regular, $y'_1(x;\Delta x) \equiv 0$.

Calculations become rather complicated for $r > 2$. Same remark as previously for general iteration functions.

5.2 Iteration functions of Newton and Schröder

They can be obtained from a Taylor series expansion of the reciprocal function $x = h(z)$ of $z = f(x)$, for a zero x of f, around the point x_i. One associates with this expansion the general iteration function

(104) $x_{i+1} = x_i + \sum_{q=1}^{r-1} \frac{1}{q!} x_z^{(q)} \, (f(x_i); \underbrace{-f(x_i), \ldots, -f(x_i)}_{q \text{ times}}),$

where the successive derivatives of x with regard to z are given implicitly (see Appendix 1). Convergence is of order r (if A is regular).

One successively has:

$r = 2$ (Newton's method),

(105)

$f'_x(x_i; x_{i+1}) = f'_x(x_i;x_i) - f(x_i)$;

$r = 3$,

(106) $$\begin{cases} f'_x(x_i; x_{i+1}) = f'_x(x_i;x_i) - f(x_i) - \frac{1}{2} f''_x(x_i;u_i,u_i), \\ f'_x(x_i;u_i) = - f(x_i); \end{cases}$$

$r = 4$,

(107) $$\begin{cases} f'_x(x_i;x_{i+1}) = f'_x(x_i;x_i) - f(x_i) - \frac{1}{2} f''_x(x_i;u_i,u_i), \\ \qquad - \frac{1}{2} f''_x(x_i;u_i,v_i) - \frac{1}{6} f'''_x(x_i;u_i,u_i,u_i), \\ f'_x(x_i;v_i) = - f''_x(x_i;u_i,u_i), \\ f'_x(x_i;u_i) = - f(x_i); \end{cases}$$

etc.

Let us prove that convergence is of order r, at least for the values 2 and 3. This is immediate for the first one, since

$$(\phi_2)'_x(x,y = x;\Delta x) = f''_{xx}(x;y = x,\Delta x) - f''_{xx}(x;x,\Delta x)$$

$$- f'_x(x;\Delta x) + f'_x(x;\Delta x) \equiv 0.$$

One has, for the second one, after simplification of notation

$$(\phi_3)'_x = f''(x;y=x,\Delta x) - f''(x;x,\Delta x) + \frac{1}{2} f'''(x;u,u,\Delta x)$$

$$+ f''(x;u,u') ,$$

and

$$(\phi_3)''_{xx} = f'''(x;y=x,\Delta x,\Delta x) - f'''(x;x,\Delta x,\Delta x) - f''(x;\Delta x,\Delta x)$$

$$+ \frac{1}{2} f^{(4)}(x;u,u,\Delta x,\Delta x) + 2f'''(x;u,u',\Delta x)$$

$$+ f''(x;u',u') + f''(x;u,u'').$$

This is identically zero, since $u = 0$ at the solution if A is regular, and since, taking (106) into account,

$$f''(x;u,\Delta x) + f'(x;u') = - f'(x;\Delta x).$$

This means (still assuming A to be regular)

$$u'(x;\Delta x) = - \Delta x .$$

Calculations are complicated but similar for $r = 4$ or larger.

Consider as an example

$$f = x^2 - a.$$

One has the iterations:

$r = 2$,

$$x_i\, x_{i+1} + x_{i+1}\, x_i = x_i^2 + a;$$

$r = 3$,

$$x_i\, x_{i+1} + x_{i+1}\, x_i = x_i^2 + a - u_i^2,$$

$$x_i\, u_i + u_i\, x_i = - x_i^2 + a.$$

5.3 Other iteration functions

We will give, merely in order to show the way of transcription, some scalar iterations drawn from J.F. Traub [5], and the corresponding matricial iterations.

Applying the first theorem of section 5.1, one has:

Scalar case,

$$u = \frac{f(x)}{f'(x)},$$

$$\Phi_1(x) = \Phi_2(x) = x - u,$$

$$\Phi_3(x) = \Phi_1(\Phi_2(x)) = x - u - \frac{f(x-u)}{f'(x-u)};$$

Matricial case,

(108) $\begin{cases} f'(x_i - u_i : x_{i+1}) = f'(x_i - u_i; x_i - u_i) - f(x_i - u_i), \\ f'(x_i ; u_i) = f(x_i). \end{cases}$

The order of convergence is four.

Applying the second theorem of section 5.1, one has:

Scalar case,

$$u = \frac{f(x)}{f'(x)} ,$$

$$\phi(x) = x - u - \frac{f(x-u)}{f'(x)} ;$$

Matricial case,

(109)
$$\begin{cases} f'(x_i; x_{i+1}) = f'(x_i; x_i - u_i) - f(x_i - u_i), \\ f'(x_i; u_i) = f(x_i) . \end{cases}$$

The order of convergence is three.

Let us give another arbitrary example:

Scalar case,

$$u = \frac{f(x)}{f'(x)} ,$$

$$\phi(x) = x - \frac{1}{6}\left[u + \frac{f(x)}{f'(x-u)} + \frac{4\, f(x)}{f'\{x - \frac{1}{4}[u + \frac{f(x)}{f'(x-u)}]\}}\right] ;$$

Matricial case,

(110)
$$\begin{cases} f'(x_i - \frac{1}{4} u_i - \frac{1}{4} v_i; x_{i+1}) = f'(x_i - \frac{1}{4} u_i - \frac{1}{4} v_i; \\ \qquad\qquad x_i - \frac{1}{6} u_i - \frac{1}{6} v_i) - \frac{4}{6} f(x_i), \\ f'(x_i - u_i; v_i) = f(x_i), \\ f'(x_i; u_i) = f(x_i). \end{cases}$$

The order of convergence is four.

5.4 Multiple roots

One obviously has first the equivalent of Schröder's method for multiple roots. It is transcribed, for instance in the case of Newton's method, as follows:

(111) $\qquad f'(x_i; x_{i+1}) = f'(x_i; x_i) - \mu f(x_i),$

where μ is the multiplicity of the root. Convergence is quadratic.

But one also might consider, as J.F. Traub [5] does, functions having the multiple root of $f(x) = 0$, say $x_{(0)}$, as a simple root and apply to them the iteration functions of sections 5.2 and 5.3. Such is the case for the function $h(x)$, defined by

(112) $\qquad f'(x; h(x)) + f(x) = 0.$

Let us differentiate (112) $\mu-1$ times. One will have:

$$f^{(\mu)}(x; h(x), \Delta x, \ldots, \Delta x) + (\mu-1) f^{(\mu-1)}(x; h'(x; \Delta x), \Delta x, \ldots, \Delta x)$$

$$+ \ldots + f'(x; h^{(\mu-1)}(x; \Delta x, \ldots, \Delta x)) + f^{(\mu-1)}(x; \Delta x, \ldots, \Delta x) \equiv 0;$$

or, setting $x = x_{(0)}$ and accounting for (72):

$$f^{(\mu)}(x_{(0)}; h(x_{(0)}), \Delta x, \ldots, \Delta x) \equiv 0.$$

If the matrix A corresponding to $f^{(\mu)}$ is regular, one infers

(113) $$h(x_{(0)}) = 0.$$

Differentiating once more, and taking (113) into account, one will get

$$\mu f^{(\mu)}(x_{(0)}; h'_x(x_{(0)}; \Delta x), \Delta x, \ldots, \Delta x) + f^{(\mu)}(x_{(0)}; \Delta x, \ldots, \Delta x) \equiv 0,$$

which shows that $h'(x_{(0)}; \Delta x)$ is not identically zero, and thus that $x_{(0)}$ is indeed a simple root of $h(x) = 0$.

Let us then apply, for instance, Newton's method to $f(x)$. One will have the iteration

(114) $$h'(x_i; x_{i+1}) = h'(x_i; x_i) - h(x_i),$$

with

(114) $$f'(x_i; h(x_i)) = -f(x_i),$$

i.e.:

(114') $$\begin{aligned} f'(x_i; x_{i+1}) + f''(x_i; h(x_i), x_{i+1}) \\ = f'(x_i; x_i) + f''(x_i; h(x_i), x_i) - f(x_i), \\ f'(x_i; h(x_i)) = -f(x_i). \end{aligned}$$

Convergence is quadratic. The order of the multiplicity does not interfere. This is an advantage over (111).

APPENDIX 1

Fréchet differentials

We will restrict ourselves in this section to explicit the symbolism which is utilized and to exhibit the Fréchet differentials of some matricial functions, especially of pseudo-inverses.

We will call the function $f(x)$, defined on a complete normed vector space and taking its values in another complete normed vector space, Fréchet-differentiable in x if we have

(115) $\quad \Delta f = f(x+\Delta x) - f(x) = f'(x;\Delta x) + \eta(x,\Delta x) \, ||\Delta x||,$

where the domain is normed by $||.||$, and where η is such that for any given number $\varepsilon > 0$, there exists a number $\theta > 0$ which verifies $||\eta|| < \varepsilon$ for $||\Delta x|| < \theta$. In other words $||\eta|| \to 0$ when $||\Delta x|| \to 0$. The function $f'(x;dx)$, which is linear and continuous in dx, is the differential of $f(x)$, and one writes

(115') $\quad df = f'(x;dx).$

Consider as an example $f(x) = x^m$. One has:

$$f'(x;dx) = x^{m-1} \cdot dx + x^{m-2} \cdot dx \cdot x + \ldots + dx \cdot x^{m-1}.$$

One will similarly write

(116) $\quad d^2 f = f''(x;dx,dx) + f'(x;d^2 x),$

where $f''(x;dx,dx)$ is the second differential of f in x, homogeneous and of degree 2 in dx ($f'''(x;u,v) = f'''(x;v,u)$), and hence symmetric in u and v), and where $d^2x = 0$, and accordingly $f'(x;d^2x) \equiv 0$, if x is an independent variable.

One easily generalizes to differentials of order r, to products of functions, and to the case of several variables. When f is a composite function, say $f(x(t))$, one will write:

$$df(x(t)) = df(t) = f'_x(x(t);x'_t(t;dt)),$$

(117) $\quad d^2f(t) = f''_x(x(t);x'_t(t;dt), x'_t(t;dt))$

$$+ f'_x(x(t); x''_t(t;dt,dt) + x'_t(t;d^2t),$$

etc.

Let $z = f(x)$ be a function involving x, and $x = h(z)$ a reciprocal function. One will have the identity $z \equiv f(h(z))$, which we will write in the form

$$z \equiv f(x(z)).$$

By applying (117) and considering an independent increment Δz, one gets:

$$\Delta z \equiv f'_x(x; x'_z(z; \Delta z))$$

$$0 \equiv f''_x(x; x'_z(z; \Delta z), x'_z(z; \Delta z))$$

$$+ f'_x(x; x''_z(z; \Delta z, \Delta z))$$

(118)

$$0 \equiv f'''_x(x; x'_z(z; \Delta z), x'_z(z; \Delta z), x'_z(z; \Delta z))$$

$$+ 3f''_x(x; x'_z(z; \Delta z), x''_z(z; \Delta z, \Delta z))$$

$$+ f'_x(x; x'''_z(z; \Delta z, \Delta z, \Delta z)),$$

and, after simplification of notation,

(118)
$$0 \equiv f_x^{(4)}(x; x', x', x', x') + 6f'''_x(x; x', x', x'') + 4f''_x(x; x', x''')$$
$$+ 3f''_x(x; x'', x'') + f'_x(x; x^{(4)}),$$

etc.

Let finally $f(x)$ be a matricial function involving the matricial variable x and an increment th of x where h is a matrix and t a scalar. The function $f(x + th)$ is a matricial function of the scalar variable t. One will be able to write for every element, and hence for the matrix, the Taylor series expansion

$$f(t) = f(0) + t\, f'_t(0) + \frac{t^2}{2!} f''_t(0) + \dots\,.$$

Setting $t = 1$ and reverting to $f(x)$ one obtains

(119) $$f(x+h) = f(x) + f'_x(x;h) + \frac{1}{2!} f''_x(x;h,h) + \ldots .$$

Some of the usual remainder formulas cannot be transposed immediately into matrix form. If $f(x)$ is a polynomial of degree ν, $f_x^{(\nu+1)} \equiv 0$, and the expansion is finite.

Let us end with the differentiation of the pseudo-inverse of a matrix $A(x)$. We will assume this latter to lie in a continuity domain (see section 3.2). We will first state a general relation valid for any inverse of an inverse semi-group (see A. Korganoff and M. Pavel-Parvu [6]), i.e. verifying the relation

(120) $$AA^{-1} A = A.$$

One gets by differentiation (with a simplified notation)

$$A (A^{-1})' A = A' - A' A^{-1} A - AA^{-1} A',$$

hence

(121) $$(A^{-1})' = - A^{-1} A' A^{-1} + u - A^{-1} A u AA^{-1},$$

where u is an arbitrary matrix involving x and Δx, having the same order as A^{-1}. Here the compatibility condition

$$- A \bar{A}^{-1} A' A^{-1} A = A' - A' A^{-1} A - A A^{-1} A',$$

i.e.

$$(I - A A^{-1}) A' (I - A^{-1} A) = 0$$

has been used. In order to check it, we write A as

(122) $$A = a\gamma^{-1} b,$$

where a and b are bases of the column and row spaces of A, respectively, and where γ is a non-singular matrix having the same rank as A. We then differentiate (122) which yields:

(123) $$A' = a'\gamma^{-1}b + a(\gamma^{-1})'b + a\gamma^{-1}b'$$
$$= a'a^{-1}A - Ab^{-1}\gamma'a^{-1}A + Ab^{-1}b'.$$

Let us now study the pseudo-inverse of A. One will have:

(124) $$A^{-1} = b^{-1} \gamma a^{-1},$$

whence, by differentiation:

(125) $$(A^{-1})' = (b^{-1})' \gamma a^{-1} + b^{-1} \gamma' a^{-1} + b^{-1} \gamma(a^{-1})'.$$

Now
$$b^{-1} = b^*(bb^*)^{-1},$$

and, bb^* being square non-singular,

$$(b^{-1})' = (b^*)' (bb^*)^{-1} - b^* (bb^*)^{-1} (bb^*)' (bb^*)^{-1}$$

$$= (b^*)' (bb^*)^{-1} - b^{-1} b' b^{-1} - b^{-1} b (b^*)' (bb^*)^{-1}$$

$$= - b^{-1} b' b^{-1} + (I - b^{-1} b) (b^*)' (b^*)^{-1} b^{-1},$$

which is indeed of the same form as (121). One would similarly obtain:

$$(a^{-1})' = - a^{-1} a' a^{-1} + a^{-1} (a^*)^{-1} (a^*)' (I - aa^{-1}).$$

Substitution into (125) yields:

$$(A^{-1})' = - b^{-1} b' A^{-1} + (I - b^{-1} b)(b^*)'(b^*)^{-1} A^{-1} + b^{-1} \gamma' a^{-1}$$

$$- A^{-1} a' a^{-1} + A^{-1} (a^*)^{-1} (a^*)'(I - aa^{-1}),$$

and, according to (123),

$$(A^{-1})' = - A^{-1} A' A^{-1} + (I - A^{-1} A)(b^*)'(b^*)^{-1} A^{-1} + A^{-1}(a^*)^{-1}(a^*)'(I - AA^{-1}).$$

If one compares to (121), the matrix u has, in this case, the form:

$$u = (b^*)'(b^*)^{-1} A^{-1} + A^{-1} (a^*)^{-1} (a^*)'.$$

APPENDIX 2

Convergence of iterative methods for solving singular linear systems by matrix splitting

The results of this section are essentially contained in the article of H.B. Keller [7]. We will deduce them again utilizing pseudo-inverses.

Let then

(127) $$Ax = c$$

be a linear system, A a singular square matrix, x and c column vectors. Consider an iterative method

(128) $$x^{(i+1)} = M^{-1}N x^{(i)} + M^{-1}c$$

associated with the splitting $A = M - N$ of A, where M is square non-singular.

One infers from (128):

(129)
$$x^{(i+1)} = (I-M^{-1}A) x^{(i)} + M^{-1}c$$
$$= (I-M^{-1}A)^{i+1} x^{(0)} + \left[(I-M^{-1}A)^{i} +...+ (I-M^{-1}A)+I\right] M^{-1}c.$$

If equation (127) is compatible, i.e. if $AA^{-1}c = c$, A^{-1} being the pseudo-inverse of A (see A. Korganoff and M. Pavel-Parvu [6]), then

$$M^{-1}c = M^{-1}A A^{-1}c = \left[I-(I-M^{-1}A)\right] A^{-1}c,$$

and (129) becomes

$$x^{(i+1)} = (I-M^{-1}A)^{i+1} x^{(0)} + \left[I - (I-M^{-1}A)^{i+1}\right] A^{-1}c$$
(130)
$$= A^{-1}c + (I-M^{-1}A)^{i+1} (x^{(0)} - A^{-1}c).$$

The matrix A being singular, $I - M^{-1}A$ necessarily has 1 as eigenvalue(s). Let us then denote by I and λ_2 the diagonal matrices of proper values of $I - M^{-1}A$, and by v_1 and v_2 the rectangular matrices of the corresponding eigenvectors. One has

$$(I - M^{-1}A) v_1 = v_1,$$
(131)
$$(I - M^{-1}A) v_2 = v_2 \lambda_2,$$

whence one infers:

$$v_1 = (I - A^{-1} A)\bar{\omega},$$
(132)
$$v_2 (I - \lambda_2) = M^{-1}A v_2,$$

$\bar{\omega}$ being an arbitrary matrix.

Let us assign $x^{(0)}$, $M^{-1}c$ and $A^{-1}c$ to the bases defined by the columns of v_1 and v_2, respectively. One will have:

$$x^{(0)} = v_1 \sigma_1 + v_2 \sigma_2,$$

$$M^{-1} c = v_1 \theta_1 + v_2 \theta_2,$$

$$A^{-1} c = v_1 \xi_1 + v_2 \xi_2.$$

Let us join (133) to (129) and (130), respectively. One obtains, according to (131):

(134)
$$x^{(i+1)} = v_1 \sigma_1 + v_2 \lambda_2^{i+1} \sigma_2 + (i+1) v_1 \theta_1 \\ + v_2 (\lambda_2^i + \ldots + \lambda_2 + I) \theta_2,$$

or

(135) $$x^{(i+1)} = v_2 \xi_2 + v_1 \sigma_1 + v_2 \lambda_2^{i+1} \sigma_2 - v_2 \lambda_2^{i+1} \xi_2.$$

If the moduli of the elements of λ_2, i.e. the eigenvalues of $M^{-1} N$ different from one, are strictly less than one, $x^{(i+1)}$ tends to

(136) $$x = v_1 \sigma_1 + \lim_{i \to \infty} (i+1) v_1 \theta_1 + v_2 (I - \lambda_2)^{-1} \theta_2$$

for (134), and to

(137) $$x = v_2 \xi_2 + v_1 \sigma_1 = A^{-1} c + (I - A^{-1} A) \bar{\omega} (\sigma_1 - \xi_1)$$

for (135).

When (127) is compatible, $x^{(i+1)}$ thus converges to (137) which of course is a solution of (127). The general term $v_1(\sigma_1-\xi_1)$ is the projection of $x^{(0)} - A^{-1}c$ on the column space of v_1 (which is the column space of $I - A^{-1}A$) parallel to the column space of v_2. This projection, however, will be expressed in terms of $A^{-1}c$, unless the column space of v_2 is orthogonal to the column space of v_1, i.e. is the column space of A^*. $x^{(0)}$ will then be able to be arbitrary in this latter space. As by virtue of (132)

$$(138) \qquad v_1^* M v_2 = 0,$$

one will have convergence to the best solution $A^{-1}c$, starting from an arbitrary initial value $x^{(0)}$ of the form $A^*\eta$ (for instance $\eta = 0$ or 1), if in particular

$$(139) \qquad M = \xi I,$$

ξ being a scalar. In order to have convergence, one should have, denoting by λ_A the non-zero eigenvalues of A,

$$(140) \qquad |1 - \xi^{-1} \lambda_A| < 1,$$

hence, assuming ξ real,

$$(140') \qquad 0 < \xi^{-1} < 2 \min \frac{Re(\lambda_A)}{|\lambda_A|^2},$$

and, if the eigenvalues λ_A are real,

$$(140'') \qquad 0 < \xi^{-1} < \frac{2}{max|\lambda_A|}.$$

When (127) is not compatible x tends linearly to infinity, except for $\theta_1 = 0$. Let us notice that $Ax^{(i+1)}$ tends, according to (131) - (133), to

$$
(141) \quad Ax = Av_2(I - \lambda_2)^{-1}\theta_2 = MM^{-1}A\, v_2(I-\lambda_2)^{-1}\theta_2 = M\, v_2\, \theta_2
$$
$$
= M(M^{-1}c - v_1\theta_1) = c - M(I-A^{-1}A)\bar{\omega}\theta_1.
$$

Hence in general (136) is a solution of (141), and of (127) only if $\theta_1 = 0$.

REFERENCES

1. F.R. Gantmacher: The theory of matrices. Vol. 1 (English translation). Chelsea Publishing Company, New York, 1959, 374 p.

2. W.T. Reid: Riccati matrix differential equations and non-oscillation criteria for associated linear differential systems. Pacific J. Math., vol. 13, 1963, pp. 665-685.

3. W.T. Reid: A matrix equation related to a non-oscillation criterion and Liapunov stability. Quart. of Appl.Math.,vol.23, 1965, pp.83-87.

4. J.E. Potter: Matrix quadratic solutions. J.SIAM Appl. Math., vol. 14, 1966, pp.496-501.

5. J.F. Traub: Iterative methods for the solution of equations. Prentice-Hall. Inc., Englewood Cliffs, New Jersey, 1964, 310 p.

6. A. Korganoff et M. Pavel-Parvu. Eléments de théorie des matrices carrées et rectangles en analyse numérique. Dunod, Paris, 1967, 441 p.

7. H.B. Keller: On the solution of singular and semi-definite linear systems by iteration. J. SIAM Numer. Anal.,Ser.B, vol.2, no. 2, 1965, pp.281-290.

8. P. Laasonen: On the iterative solution of the matrix equation $AX^2 - I = 0$. MTAC, vol. 12, no 62, 1958, pp. 109-116.

Dr. M. Pavel-Parvu
and
Dr. A. Korganoff
Cie Bull-General Electric
94 avenue Gambetta
Paris 20^e

H. Rutishauser

Zur Problematik der Nullstellenbestimmung bei Polynomen

1. Wenn wir die Nullstellen eines Polynoms bestimmen wollen, müssen wir uns vorerst einige Fragen stellen, die damit zusammenhängen, dass Polynome in praktischen Problemen selten als Selbstzweck, sondern meist nur als (oft untaugliches) Mittel zur Lösung einer umfassenderen Aufgabe auftreten. Wo dies aber der Fall ist, muss die Problemstellung "Bestimme die Nullstellen eines Polynoms" immer im Rahmen der grösseren Aufgabe betrachtet werden, da sie sonst nur allzuleicht jeden Sinn verlieren kann. Die erwähnten Fragen lauten nun:

    a) Was ist ein Polynom?

    b) Wozu brauche ich seine Nullstellen?

    c) Brauche ich sie überhaupt?

    d) Wie genau brauche ich sie?

Ein allfälliges Rechenprogramm muss sich an diesen Fragen ausrichten, denn die Schwierigkeiten der Wurzelbestimmung, die in den zahlreichen Bemühungen um das Problem zum Ausdruck kommen, sind nicht zuletzt darauf zurückzuführen, dass man sich diese Fragen oft nicht stellt.

Die Tatsache, dass viele Probleme formal sehr leicht auf die Nullstellenbestimmung zurückgeführt werden können, und damit durch eine vorfabrizierte Prozedur "Nullstellenbestimmung" prak-

tisch in nichts aufgelöst würden, hat natürlich das allgemeine
Interesse an diesem Problem enorm geweckt. Es muss aber gesagt werden, dass uns die beschränkte Genauigkeit jeder numerischen Rechnung daran hindert, eine für alle Zwecke sachgerechte und narrensichere Wurzelbestimmungsprozedur aufzustellen. Diese Schwierigkeit
ist grundsätzlicher Natur, sie kann durch Anwendung einer mehrfach
genauen Arithmetik nicht überwunden, sondern nur soweit verschoben
werden, dass sie praktisch nicht mehr in Erscheinung tritt. Man
wird sich daher darauf beschränken müssen, für die verfügbare Rechengenauigkeit das Optimum an Wurzelgenauigkeit und Rechenzeit
herauszuholen, wie dies z.B. Herr Nickel in seiner Arbeit [6] getan
hat. Bessere Resultate kann man nur dann erwarten, wenn man die
Wurzelbestimmungsmethode auf das spezielle Problem ausrichtet.

2. Beginnen wir mit der Frage a), die man in der Regel
durch Anschreiben von $P(x) = \text{SUMME}_{k=0}^{n} c_k x^k$ beantwortet zu haben
glaubt. Abgesehen davon, dass man diesem Ausdruck eine formale oder
eine arithmetische Interpretation geben kann - die sich numerisch
verschieden auswirken - ist dies ja nur eine von vielen möglichen
Darstellungsarten. Andere sind

- Entwicklung nach einem System von Orthogonalpolynomen,
also

$$P(x) = \text{SUMME}_{k=0}^{n} c_k p_k(x) ,$$

wobei die $p_k(x)$ jetzt aber durch eine unendliche Tridiagonalmatrix und nicht etwa durch eine Entwicklung nach
Potenzen von x definiert sind.
- Vorgabe der Funktionswerte von $P(x)$ an $n+1$ Stellen
$x_0, x_1, \ldots, x_n$, oder allgemeiner: Total $n+1$ Funktionswerte
und Ableitungen.

- Ausdrücke der Form (siehe H.R. Schwarz [7])

$$PRODUKT_{k=0}^{n} (x^2 + a_k^2) + cx \, PRODUKT_{k=0}^{n-1} (x^2 + b_k^2),$$

oder allgemeiner: Ganz rationale Ausdrücke $G(P_1, P_2, \ldots, P_r)$ in den Polynomen P_i.

Als Beispiel für die unterschiedlichen Voraussetzungen, die die verschiedenen Darstellungsarten bieten, betrachten wir

$$P(x) = 1 - 13.7\, x + 67.5\, x^2 - 153\, x^3 + 162\, x^4 - 64.8\, x^5.$$

Die Auswertung an der Stelle 0.7 liefert hier den Wert 0.011264 als Summe von Termen bis zur Grösse 52. Man hat also starke Auslöschung dank der Tatsache, dass trotz grosser Koeffizienten $|P(x)| < 0.031$ im Intervall $0.15 < x < 0.85$. Nun ist aber auch

$$P(x) = -0.1265625\, T_5^* - 0.3515625\, T_3^* - 0.521875\, T_1^*,$$

wobei hier die T^* die Tschebyscheff-Polynome für das Intervall $[0,1]$ sind. Auswertung dieses Ausdrucks an der Stelle 0.7 liefert

$$P(0.7) = -0.11861 + 0.331825 - 0.521875 = 0.011264,$$

die Auslöschung ist also ca. 100 mal geringer. Schliesslich kann $P(x)$ auch durch 6 Stützwerte definiert werden, beispielsweise

x :	0	0.2	0.4	0.6	0.8	1
y :	1	-0.025536	0.011648	-0.011648	0.025536	-1,

aus denen $P(x)$ mit Hilfe der Lagrange'schen Interpolationsformel leicht und genau berechnet werden kann. Man erhält z.B. für $x=0.7$:

$$P(0.7) = \text{SUMME}_{k=0}^{5} y_k L_k(0.7) = \qquad [L_k = \text{Lagrange-Funktionen}]$$
$-0.01171875\ -0.00209475\ -0.003185\ -0.009555\ +0.0104375\ +0.02734375,$

also praksich ohne jede Auslöschung. Allerdings ist die Auswertung mittels dividierter Differenzen nach Newton nicht ebenso vorteilhaft.

3. Aus der Erkenntnis, dass Polynome auf andere Weise eventuell besser dargestellt werden können, ergibt sich konsequenterweise das Bedürfnis, Wurzelbestimmungsmethoden auf diese Darstellungsarten umzuformen.

Beispielsweise erhält man für eine Darstellung $\text{SUMME } c_k p_k(x)$ bekanntlich ohne weiteres einen dem Horner-Schema entsprechenden Algorithmus für die Berechnung von $P(x)$ und $P'(x)$ zu gegebenem x (Vgl. Clenshaw [5] für die Tschebyscheff-Polynome), und damit die Möglichkeit, das Newton-Verfahren direkt mit den T-Koeffizienten durchzuführen. Ebenso lässt sich die Bauer'sche Treppeniteration direkt mit den T-Koeffizienten der Treppen-Polynome durchführen.

Entsprechendes gilt für die Darstellung von $P(x)$ durch Stützwerte, die zu einer interessanten Variante des Newton'schen Verfahrens führt. Sind nämlich $x_0, x_1, \ldots, x_n$, $y_0, y_1, \ldots, y_n$ die gegebenen Stützstellen und Stützwerte, und $w_0, w_1, \ldots, w_n$ die zugehörigen Lagrange-Gewichte, dann wird

$$P'(x_k) = -\frac{1}{w_k} \text{SUMME}_{j \neq k} \; w_j \frac{y_j - y_k}{x_j - x_k}.$$

Man könnte nun so vorgehen, dass man mit Hilfe dieser Formel die Stützwerte x_1 bis x_n durch Newton-Approximationen der Nullstellen ersetzt, während man z.B. x_0 festhält. Bei einem solchen Vorgehen würde man zwar nicht die Genauigkeit der Wurzelwerte verbessern, aber viel bessere Abbrechkriterien erhalten, da mit der fortlaufenden Annäherung der x_k an die Wurzeln des Polynoms die Stützwerte $y_1, y_2, \ldots, y_n$ auch numerisch wirklich gegen 0 streben; es bleibt somit nur der zu x_0 gehörige Lagrange-Term, der damit $P(x)$ angenähert in faktorisierter Form darstellt.

4. Besonders interessante Aspekte vermittelt die für die Lagrangesche Polynomdarstellung formulierte Bauer'sche Treppeniteration. Die Rechenvorschrift für diese lautet, wenn man die Iteration rückwärts laufen lässt (Bauer-Samelson [4]):

$$P_k^{(j-1)} := (P_{k+1}^{(j)} + q_{n-k}^{(j)} \times P_k^{(j)})/(x-z) \quad (k=0,1,\ldots,n-1; \; j=0,-1,-2,\ldots),$$

wobei $P_k^{(j)}$ immer ein Polynom $x^k + \ldots$ bedeutet und $P_n^{(j)}$ für alle j das gegebene Polynom vom Grad n (Höchstkoeffizient 1) ist, dessen Nullstellen $z_1, z_2, \ldots, z_n$ zu bestimmen sind. z ist eine beliebige Konstante, während der Skalar $q_{n-k}^{(j)}$ jeweils so zu bestimmen ist, dass die Division durch $x-z$ aufgeht. Es gilt:

Falls $|z_1-z| < |z_2-z| < \ldots < |z_n-z|$,

dann $\lim\limits_{j} q_k^{(j)} = z_k - z \quad (k=1,2,\ldots,n)$, so dass sich beispielsweise q_1+z mit fortschreitender Iteration der z am nächsten liegenden Wurzel von P_n nähert.

Da die Rechenvorschrift nichts über die Art der Polynomdarstellung aussagt, kann man es beispielsweise mit der Darstellung durch Stützwerte versuchen; es wird dann, wobei z eine feste Zahl ist:

$$q_{n-k}^{(j)} = -P_{k+1}^{(j)}(z)/P_k^{(j)}(z),$$

$$P_k^{(j-1)}(x_i) = (P_{k+1}^{(j)}(x_i) + q_{n-k}^{(j)} \times P_k^{(j)}(x_i))/(x_i-z)$$

$$(j = 0,-1,-2,\ldots\ ;\ k = 0,1,\ldots,n-1\ ;\ i = 0,1,\ldots,k).$$

Die $P_k(z)$-Werte werden nach Lagrange bestimmt; es seien $c_0, c_1, \ldots, c_n$ die Koeffizienten, mit denen $P(z) = \text{SUMME } c_k P(x_k)$. Wenn ferner $p[k,i]$ für jedes j den Funktionswert $P_k^{(j)}(x_i)$ bezeichnet, und $p0$ die Konstante $P_n(z)$ bedeutet, erhält man folgendes ALGOL-Programm:

```
za : t := p0 ;
    for k := n-1 step -1 until k1 do
    begin
        s := 0 ;
        for i := 0 step 1 until n do s := s + c[i]×p[k,i] ;
        q[n-k] := -t/s ;
        t := s
    end ;
    for k := k1 step 1 until n-1 do
        for i := 0 step 1 until n do
            p[k,i] := (p[k+1,i] + q[n-k]×p[k,i])/(x[i]-z) ;
    goto za ;
```

Als triviales Beispiel wird die Berechnung der ersten Nullstelle der Besselfunktion $J_0(x)$ aus Funktionswerten an den Stellen 2.2, 2.3, 2.4, 2.5, 2.6, 2.7 gezeigt. Da diese offensichtlich im Intervall $[2.4, 2.5]$ liegt, wählt man $z = 2.45$. Im übrigen genügt es in diesem Beispiel, mit P_4 und P_5 zu arbeiten; das obige Programm wird also mit $n=5$, $k1=4$ benützt:

x_k	$P_5(x_k)=J_0(x_k)$	$P_4^{(0)}$	$P_4^{(1)}$	$P_4^{(2)}$	$P_4^{(3)}$	$P_4^{(4)}$
2.2	.1103623	1	−.5343564	−.5389745	−.5388091	−.5388111
2.3	.0555398	1	−.5251106	−.5299953	−.5298285	−.5298306
2.4	.0025077	1	−.5146900	−.5198344	−.5196663	−.5196683
2.5	−.0483838	1	−.5031400	−.5085360	−.5083683	−.5083693
2.6	−.0968050	1	−.4905213	−.4961582	−.4959906	−.4959922
2.7	−.1424494	1	−.4768904	−.4827604	−.4825921	−.4825940
2.45	−.0232268	1	−.5090526	−.5143240	−.5141559	−.5141574
q-Werte			−.0456275	−.0451598	−.0451746	−.0451745

Schliesslich : Nullstelle $= z + q = 2.4048255$

5. Nun zur Frage b) betreffend den Zweck, dem die Nullstellen dienen sollen: Es ist ja oft so, dass man eigentlich Pole einer rationalen Funktion sucht und diese als Nullstellen des Nenners berechnet, oder Eigenwerte einer Matrix als Nullstellen des charakteristischen Polynoms. Dass die Eigenwertbestimmung auf diesem Weg unzweckmässig ist, ist nun doch allgemein bekannt, aber auch die Bestimmung der Pole einer rationalen Funktion als Nullstellen des Nennerpolynoms ist im Grunde genommen unsachgemäss. Tatsächlich ist den Polen einer rationalen Funktion ein Residuum zugeordnet; dies ist eine wertvolle zusätzliche Information, die verloren geht, wenn man nur mit dem Nenner arbeitet. Beispielsweise hat die rationale Funktion

$$R(x) = \frac{1.000001x - 1.001001}{x^2 - 2.001x + 1.001}$$

die Pole 1 und 1.001 mit den zugehörigen Residuen 1 und 10^{-6}. Offensichtlich wird der erste Pol durch die Daten viel genauer bestimmt als der zweite und dürfte in der Praxis auch viel wichtiger sein. Ein Nullstellenprogramm für Polynome kann diese Unterscheidung nicht machen, sondern liefert beide Pole ungefähr gleich ungenau.

Als Konsequenz dieser Erkenntnis ergibt sich, dass man Wurzelbestimmungsmethoden sinngemäss auf andere Datenformen (rationale Funktionen, Fourierreihen, Matrizen, etc.) umformulieren sollte, statt Probleme um jeden Preis auf algebraische Gleichungen zurückzuführen. Natürlich ist schon viel in dieser Richtung getan worden; beispielsweise ist die LR-Transformation im wesentlichen die auf Matrizen angewendete Treppeniteration, und auch das Graeffe-Verfahren ist schon für Matrizen formuliert worden. Man beachte ferner die Arbeiten von W. Specht [8] über die Lage der Nullstellen eines nach Orthogonalpolynomen $p_k(x)$ entwickelten Polynoms $P(x)$.

6. Dagegen scheint, dass für die Bestimmung der Pole einer rationalen Funktion bisher wenig getan wurde, wenn man davon absieht, dass die Bernoulli-Lagrange'sche Methode und die Treppeniteration eigentlich Polbestimmungsmethoden sind. Bekannt ist ferner die Newton-Modifikation

$$x_{k+1} = x_k + \frac{f(x_k)}{f'(x_k)},$$

die aber lediglich den Spezialfall $p=-1$ der Newtonformel für eine p-fache Nullstelle darstellt. Auf eine rationale Funktion $1/P(x)$ angewendet bringt diese Methode nichts neues; sie reduziert sich dann auf das Newton-Verfahren. Sonst aber sticht sie in vorteilhafter Weise von diesem ab, indem sie beispielsweise Polhaufen mit nicht-verschwindender Residuensumme aus der Ferne wie einen einfachen Pol ansieht. Beispielsweise erhält man für die oben erwähnte rationale Funktion $R(x)$ mit den Polen 1 und 1.001 aus $x_0 = 0.9$ sofort $x_1 = 1 + 10^{-9}$. Für den allgemeinen Fall bedeutet dies, dass nur die Pole mit grossen Residuen praktisch existent sind, was die Aufgabe nur erleichtert. Die schwachen Pole kommen erst nach Subtraktion der zu den starken Polen gehörigen Hauptteile zum Vorschein und können dann ebenfalls bestimmt werden.

Bemerkenswert, wenn auch wenig beachtet, ist das Gegenstück des Graeffe-Verfahrens für rationale Funktionen, welches wie folgt beschrieben werden kann: Sei $R_0(x) = R(x)$ die gegebene rationale Funktion. Man bildet für $k = 0,1,2,...$

$$R_{k+1}(x) = \frac{1}{2} (R_k(\sqrt{x}) + R_k(-\sqrt{x})).$$

Mit wachsendem k zerfällt $R_k(x)$ in Teilbrüche, von denen jeder Pole gleichen Absolutbetrages hat. Durch Wiederholung des Verfahrens für $x R(x)$, $x^2 R(x)$, etc., erhält man schliesslich die vollständige Partialbruchzerlegung von $R(x)$, ohne dass man jemals eine 2^k-te Wurzel zu berechnen hätte. Natürlich ist auch dies nichts Neues; denn genau dieser Gedanke könnte dem Verfahren von Brodetzky-Smeal zugrundegelegt werden, indem man $R(x) = P'(x)/P(x)$ setzt. Es soll aber nicht verschwiegen werden, dass dem so erweiterten Graeffe-Verfahren immer noch einige (aber nicht mehr alle!) der bekannten Mängel anhaften.

7. Nun zur Frage c), ob man die Nullstellen wirklich brauche. Die Frage mag reichlich dumm tönen, sie muss aber tatsächlich gelegentlich verneint werden. Wenn es sich z.B. darum handelt, für die Lösung einer homogenen linearen Differentialgleichung

$$SUMME_{k=0}^{n}\, a_k y^{(k)}(x) = 0 \qquad \text{(mit konstanten } a_k\text{)}$$

eine *exakte* Differenzengleichung

$$SUMME_{k=0}^{n}\, b_k y(t+k\times h) = 0$$

aufzustellen, so ist dies äquivalent mit der Aufgabe, zu einem Polynom $SUMME\ a_k x^k$ mit den Wurzeln z_j ein Polynom $Q(x) = SUMME\ b_k x^k$ mit den Wurzeln $exp(h \times z_j)$ zu berechnen. Es ist zwar naheliegend, zunächst die z_j zu bestimmen, und dann $Q(x)$ aus seinen Linearfaktoren aufzubauen. Tatsächlich braucht man aber die Wurzeln z_j überhaupt nicht, denn die b_k lassen sich direkt aus den a_k berechnen. Hierzu dient das Graeffe-Verfahren, das für diesen Zweck so modifiziert wird, dass ein Schritt ein Polynom mit den Wurzeln z_j in ein Polynom mit den Wurzeln $z_j + r\, z_j^2$ umwandelt. Dieser Algorithmus wird dann (beispielsweise für 11-stellige Genauigkeit) nacheinander mit $r = 2^{-12}/3,\ 2^{-12}/3,\ 2^{-12}/3,\ 2^{-12},\ 2^{-11},\ 2^{-10},\ldots,\ 2^{-3},\ 2^{-2}$ angewendet.

Auch das Filterproblem enthält eine ähnliche Problemstellung als Teilaufgabe, nämlich aus einem Polynom mit den nicht negativ-reellen Nullstellen z_j ein Polynom mit den Nullstellen $\sqrt{z_j}$ (deren Realteile positiv sein sollen) zu bestimmen. Auch hier wird man eine Lösungsmöglichkeit vermuten, die die z_j nicht benützt. Tatsächlich wurde hierzu schon vor längerer Zeit von F.L. Bauer [3] die Treppeniteration vorgeschlagen; es hat sich später gezeigt, dass die in [2] verwendete Cayley-Parametrisierung vermieden werden kann, wenn man als Startpolynom die modulo $P(x^2)$ auf ein Polynom reduzierte Funktion e^{cx} (mit passendem c) benützt, und auch immer mit diesem Polynom iteriert, bzw. im Sinne der abgekürzten Treppeniteration (F.L. Bauer [1]) laufend quadriert. Freilich ist die Rolle der Rundungsfehler bei diesem Vorgehen noch nicht geklärt.

8. Zur Frage d) betr. die benötigte Genauigkeit ist vorerst einmal zu bemerken, dass man - von speziellen Massnahmen, wie exakte Faktorzerlegung eines Polynoms mit ganzzahligen Koeffizienten abgesehen - mehrfache oder fast mehrfache Nullstellen grundsätzlich nicht mit derselben Genauigkeit bestimmen kann, wie gerechnet wurde. Wir haben mit dieser Tatsache zu leben, und es scheint mir wichtig, dass von derart ungenau berechneten Wurzeln - ob man für ihre Fehler Schranken habe oder nicht - der bestmögliche Gebrauch gemacht wird. Immerhin sei festgestellt, dass wenn wir auch zulassen müssen, dass mehrfache Wurzeln ungenau bestimmt werden, dennoch verlangt werden kann, dass ihre Fehler in solcher Weise korreliert sind, dass beim Zusammenmultiplizieren der Linearfaktoren das ursprüngliche Polynom innerhalb der Maschinengenauigkeit reproduziert wird. Man würde also (bei einer Rechengenauigkeit von beispielsweise *1:5000*) *1.0007* und *1.0009* nicht als Nullstellen des Polynoms x^2-2x+1 akzeptieren, müsste aber die an sich ungenaueren Werte *0.99* und *1.01* zulassen. Man kann diese Korrelation der Rechenfehler erzwingen, indem man das Abdividieren der laufend berechneten Linearfaktoren nicht vom Höchstkoeffizienten aus, sondern von beiden Seiten her vornimmt und dann die beiden Teile an einer passenden Stelle zusammenflickt.

Als primitives Beispiel für die Verwendung ungenauer Wurzeln betrachten wir die Differentialgleichung

$$y'' - 2y' + y = 0, \quad \text{mit } y(0) = 1, \quad y'(0) = 0 \;.$$

Die charakteristische Gleichung $a^2-2a+1=0$ habe die Wurzeln $a=1.01$ und 0.99 geliefert, woraus sich als Lösung

$$y = 50.5\, e^{0.99x} - 49.5\, e^{1.01x} \quad [\text{anstelle von } (1-x)e^x]$$

ergibt. Das liefert für kleines x tatsächlich ein der Rechengenauigkeit (*1:5000*) entsprechendes Resultat; selbst für $x=2$ erhält man noch -7.388563 statt -7.389056, aber das asymptotische Verhalten wird erheblich verfälscht.

Würde man die Wurzeln mit der vollen Rechengenauigkeit (*1:5000*), hier also etwa als *1.0001* und *0.9999* bestimmt haben, hätte man

$$y = 5000.5\, e^{0.9999x} - 4999.5\, e^{1.0001x}$$

erhalten, was wohl theoretisch genauer ist, aber bei *4*-stelliger Rechnung für kleines x infolge Auslöschung überhaupt kein Resultat mehr liefert. Wir werden hier also mit der Tatsache konfrontiert, dass man Nullstellen auch *zu genau* berechnen kann, und es scheint sogar, dass die Ungenauigkeit, die den numerisch bestimmten Nullstellen normalerweise anhaftet, gerade irgendwie optimal ist.

Nun kann man allerdings die Auslöschung bei der y-Berechnung im obigen Beispiel durch Verwendung der hyperbolischen Funktionen weitgehend unterbinden, so dass unser Argument betreffend die zu genaue Wurzelbestimmung dahinfällt. Man erhält nämlich

$$y = e^x \left(Ch(dx) - \frac{Sh(dx)}{d} \right), \quad \text{mit} \quad d = 0.01 \quad \text{bzw.} \quad 0.0001.$$

Im allgemeinen Fall, wo p Wurzeln $z_1, z_2, \ldots, z_p$ nahe beieinanderliegen, ist dies allerdings sehr kompliziert. Man kann dann stabile Darstellungen für y mit Hilfe der ganz transzendenten, in den Wurzeln symmetrischen Funktion

$$E(x) = \text{SUMME}_{k=1}^{p} w_k exp(z_k x) \qquad [w_k = \text{Lagrange-Gewichte für die } z_k]$$

und ihrer Ableitungen, erhalten, sofern man sich nicht einfach mit Potenzreihenentwicklungen begnügen will.

Als Konsequenz der hier genannten Tatsachen folgt offensichtlich die Notwendigkeit, die Abschätzung der Wurzelgenauigkeit auf die spätere Verwendung dieser Wurzeln auszurichten, was in zahlreichen Fällen bedeutet, dass auch die Korrelationen der Fehler der einzelnen Wurzeln abgeschätzt werden müssen.

LITERATUR

1. F.L. Bauer, Beiträge zur Entwicklung numerischer Verfahren für programmgesteuerte Rechenanlagen. I. Quadratisch konvergente Durchführung der Bernoulli-Jacobi'schen Methode. Sitzungsberichte der Bayrischen Akademie der Wissenschaften, 1954, Mathematisch-Naturwissenschaftliche Klasse, p. 275-303.

2. F.L. Bauer, Ein direktes Iterationsverfahren zur Hurwitz-Zerlegung eines Polynoms. Arch. Elektr. Uebertragung 9 (1955) p. 285-290.

3. F.L. Bauer, Beiträge zur Entwicklung numerischer Verfahren für programmgesteuerte Rechenanlagen. II. Direkte Faktorisierung eines Polynoms. Sitzungsberichte der Bayrischen Akademie der Wissenschaften, 1956, Mathematisch-Naturwissenschaftliche Klasse, p. 163-203.

4. F.L. Bauer und K. Samelson, Polynomkerne und Iterationsverfahren. Math. Zeitschr. *67* (1957) p. 93-98.

5. C.W. Clenshaw, A Note on the Summation of Chebyshev Series. M.T.A.C. *9* (1955) p. 118-120.

6. K. Nickel, Die numerische Berechnung der Wurzeln eines Polynoms. Num. Math. *9* (1966) p. 80-98.

7. H.R. Schwarz, Das Approximationsproblem für frequenzunsymmetrische Bandfilter. Mitt. Nr. 9 aus dem Institut für angew. Mathematik der ETH, p. 65-99. Birkhäuser Verlag Basel 1964.

8. W. Specht, Die Lage der Nullstellen eines Polynoms. Mathematische Nachrichten *15* (1956) p. 353-374; *16* (1957) p. 257-263; *16* (1957) p. 369-389; *21* (1960) p. 201-222.

Prof. Dr. H. Rutishauser
Eidgenössische Technische Hochschule
Zürich
Schweiz

Johann Schröder

Factorization of Polynomials by Generalized Newton Procedures

Various iterative methods for calculating roots of polynomials or splitting polynomials into factors can be considered as special cases of a generalized Newton procedure, applied to some operator equation $G\mathbf{u} = \mathbf{0}$. Such methods are, for example, different versions of Bernoulli's method, the "Treppeniteration", the ordinary method of Newton for calculating a root of a polynomial, and Bairstow's method.

There are many ways of defining an appropriate operator G (see Section 5.1). The definition of G as in Section 1 seems to have some advantages. For example, the second derivative $G''(\mathbf{x})$ is constant, i.e. independent of $\mathbf{x}$, which is useful for convergence and estimation theories (see Section 4).

From the generalized Newton procedure in Section 2, we derive here only a few known methods for splitting polynomials into factors (see Section 3).

The justification for reformulating the known methods in a new way is given by the following remarks. The unified and rather simple description may serve in better understanding the connections between the different methods, and in comparing these methods, as well as in gaining a unified convergence and estimation theory.

Moreover, it is easy to derive a series of new methods using the general idea of Section 2. For example, it may be allowed to remark, that the method of Section 3.4, proposed by K. Samelson [10], has been derived here independently from the general principle. Notice also that in most parts of this paper we do not use any assumptions on the form in which the occurring polynomials are given or calculated.

This paper has been stimulated by a paper of F.L. Bauer and K. Samelson [3], and the essential part of our paper has been written right after the publication of their work. Although, in the meantime, we have not worked very much on the problem, and have, for example, not developed a unified estimation theory, it may be worthwhile indicating some possibilities at this meeting which is particularly concerned with polynomials.

1. <u>The Problem of Factorization</u>

Let be given a polynomial $P(z)$ of the exact degree N. Denote the "leading coefficient" of $P(z)$ by c, i.e. $c = \lim_{|z| \to \infty} P(z)/z^N$. We want to split $P(z)$ into factors of a certain prescribed form:

$$P(z) = u^1(z) u^2(z) \ldots u^p(z) U^p(z)$$

with polynomials $u^k(z)$ of given exact degree $m_k \geq 1$ and leading coefficient 1, and a polynomial $U^p(z)$ of degree $M_p = N - m_1 - \ldots - m_p \geq 1$. Of course, U^p then has leading coefficient c.

This problem is equivalent to the following one:

Find polynomials $u^k (k = 1, 2, \ldots, p)$ *as described above and polynomials* $U^k (k = 1, 2, \ldots, p)$ *of degree* $M_k = N - m_1 - \ldots - m_k$ *with leading coefficient* c *such that*

$$U^{k-1} = u^k U^k \quad (k = 1, 2, \ldots, p)$$

with

$$U^0 = P.$$

This problem shall now be formulated in an abstract way. Let R be the set of "vectors"

$$\mathbf{u} = (u^1, u^2, \ldots, u^p; U^1, \ldots, U^p) = (u^k, U^k)$$

whose "coefficients" u^k and U^k are polynomials of degree at most m_k and M_k, respectively. Denote by D the subset of all $\mathbf{u} \in R$, such that the u^k have leading coefficient 1, and the U^k have leading coefficient c. Moreover, let S denote the set of vectors

$$\mathbf{s} = (s^1, \ldots, s^p) = (s^k)$$

whose coefficients s^k are polynomials of degree at most $M_k - 1$.

Then, by

$$G\mathbf{u} = (U^0 - u^1 U^1, \ldots, U^{p-1} - u^p U^p) = (U^{k-1} - u^k U^k)$$

there is defined an operator G on D which maps this set into S.

Using these terms the problem may be formulated as follows.

Find a solution $\mathbf{u}$ *(ϵD) of the equation*

(1.1) $$G\mathbf{u} = \mathbf{0}.$$

Here, $\mathbf{0}$ denotes the null element of S. Clearly, R and S are linear spaces.

For later use, we define a linear subspace L of R. L shall be the set of all $\mathbf{h} = (h^k, H^k) \epsilon R$ such that the h^k are polynomials of degree at most $m_k - 1$, and the H^k are polynomials of degree at most $M_k - 1$. Then $\mathbf{u} - \tilde{\mathbf{u}} \epsilon L$ for $\mathbf{u}, \tilde{\mathbf{u}} \epsilon D$ and $\mathbf{u} + \mathbf{h} \epsilon D$ for $\mathbf{u} \epsilon D, \mathbf{h} \epsilon L$.

2. The Generalized Newton Procedure

For $\mathbf{x} = (x^k, X^k) \epsilon D$ and $\mathbf{h} = (h^k, H^k) \epsilon L$, the difference $G(\mathbf{x}+\mathbf{h}) - G\mathbf{x}$ assumes the following form:

(2.1) $$G(\mathbf{x}+\mathbf{h}) - G\mathbf{x} = (H^{k-1} - h^k X^k - x^k H^k) - (h^k H^k)$$

with

$$H^0 = 0.$$

The first of the two summands of the right-hand side of (2.1) is homogeneous in $\mathbf{h}$ of order 1, the second summand is homogeneous of order 2. Therefore, we define the derivative $G'(\mathbf{x})$ of G in $\mathbf{x}$ by

(2.2) $$G'(\mathbf{x})\mathbf{h} = (H^{k-1} - h^k X^k - x^k H^k).$$

This is a linear operator which maps the linear subspace L into S.

In Section 4, we will define norms on R and S. With respect to these norms, $G'(\mathbf{x})$ is a Fréchet derivative.

We consider now the

Generalized Newton Procedure:

(2.3) $\quad G\mathbf{u}_n + G'(\mathbf{x}_n)(\mathbf{u}_{n+1} - \mathbf{u}_n) = \mathbf{0} \quad (n = 0,1,2,\ldots)$

with elements $\mathbf{x}_n \varepsilon D$.

The procedure for constructing these elements $\mathbf{x}_n$ need not be specified at the moment. The element $\mathbf{x}_n$ may depend on $\mathbf{u}_0, \mathbf{u}_1, \ldots, \mathbf{u}_n$. For example, with $\mathbf{x}_n = \mathbf{u}_n$, formula (2.3) represents the regular Newton method for solving equation (1.1). In the following sections, we will also use other possibilities for choosing $\mathbf{x}_n$. Generalized Newton procedures of type (2.3) have been investigated in [11].

Suppose now, that elements $\mathbf{u}_0, \mathbf{u}_1, \ldots, \mathbf{u}_n$ satisfying (2.3) with certain $\mathbf{x}_0, \ldots, \mathbf{x}_{n-1}$ have been found. Then, if $G'(\mathbf{x}_n)$ has an inverse which is defined on all of S, the approximation $\mathbf{u}_{n+1}$ is uniquely defined by (2.3).

If $\mathbf{x} = (x^k, X^k) \varepsilon L$ is such that for all $k = 1,2,\ldots,p$ the two polynomials x^k, X^k have no common roots, then $G'(\mathbf{x})$ has an inverse defined on S.

For proving the last statement, we consider the equation

$$G'(\mathbf{x})\mathbf{h} = \mathbf{s}$$

with given $\mathbf{s}\varepsilon S$, more explicitely:

$$H^{k-1} - h^k X^k - x^k H^k = s^k \quad (k = 1,2,\ldots,p).$$

We show that for fixed k and given H^{k-1}, s^k there exist polynomials h^k, H^k (of the prescribed form) which satisfy the k-th equation, and that these polynomials are uniquely determined.

For simplicity, we drop the fixed index k, and write the k-th equation as

$$hX + xH = Q \text{ with } Q = H^{k-1} - s^k.$$

Let $z_1,\ldots,z_q$ (with $q = m_k$) be the roots of x, and represent h by using Newton's interpolation formula:

$$h = \gamma_1 + \gamma_2(z-z_1) + \gamma_3(z-z_1)(z-z_2) + \ldots + \gamma_q(z-z_1)\ldots(z-z_{q-1}).$$

If $X(z_i) \neq 0$ ($i = 1,2,\ldots,q$), as assumed above, the γ_j and H are uniquely defined by the following procedure:

$$Q_{(0)}(z) = Q(z);$$

$$\gamma_{j+1} X(z_{j+1}) = Q_{(j)}(z_{j+1}),$$

$$Q_{(j+1)}(z) = (Q_{(j)}(z) - \gamma_{j+1} X(z))/(z-z_{j+1}) \quad (j = 0,1,\ldots,q-1);$$

$$H(z) = Q_{(q)}(z).$$

Theoretically, this procedure describes a possibility of calculating $\mathbf{u}_{n+1}$. In praxis however, this procedure will not be used in general.

In more detail, the Newton procedure (2.3) can be written as follows.

Generalized Newton procedure:

$$(2.4) \begin{cases} U_n^0 = P, \\ w_{n+1}^k x_n^k + x_n^k U_{n+1}^k = U_{n+1}^{k-1} - (u_n^k - x_n^k)(U_n^k - X_n^k), \\ u_{n+1}^k = x_n^k + w_{n+1}^k, \quad (k = 1,2,\ldots,p;\ n = 0,1,2,\ldots). \end{cases}$$

Note, that this procedure assumes a particularly simple form if one chooses

$$x_n^k = u_n^k, \quad \text{or} \quad X_n^k = U_n^k$$

for all k and n.

3. Some Examples

We show now that some known approximation methods are special cases of the generalized Newton procedure (2.3), (2.4).

In all examples which we will describe, the index k shall assume the values $k = 1, 2, \ldots, p$, and $n = 0, 1, 2, \ldots$. If $p = 1$, we will drop the index k and write $u_n = u_n^k$, for example.

3.1 Treppeniteration and QD-Algorithm

Choose: p arbitrarily, $m_k = 1$;

$$x_n^k(z) = z, \quad X_n^k = U_n^k .$$

Write, moreover,

(3.1) $$w_{n+1}^k(z) = \alpha_{n+1}^k, \quad u_n^k = z - z_n^k .$$

Then, (2.4) becomes the method of

Treppeniteration:

$$U_n^0 = p$$

$$\alpha_{n+1}^k \, U_n^k + z \, U_{n+1}^k = U_{n+1}^{k-1} ,$$

or

$$U_n^0 = P,$$

$$\alpha_{n+1}^k = U_{n+1}^{k-1}(0)/U_n^k(0),$$

$$U_{n+1}^k = (U_{n+1}^{k-1} - \alpha_{n+1}^k U_n^k)/z,$$

$$z_{n+1}^k = -\alpha_{n+1}^k,$$

(see [1], [2], [3]).

This method is closely connected with the QD-algorithm [9].

3.2 Bernoulli's method

For $p = 1$, the Treppeniteration becomes the Lagrange form of *Bernoulli's method*:

$$\alpha_{n+1} = P(0)/U_n(0),$$

$$U_{n+1} = (P - \alpha_{n+1} U_n)/z,$$

$$z_{n+1} = -\alpha_{n+1}.$$

It is well known that the numbers α_{n+1} can be calculated without computing the polynomials U_n (see, for example, [3]).

3.3 Newton-type Treppeniteration

For p arbitrary, $m_k = 1$,

$$x_n^k = u_n^k, \quad X_n^k = U_n^k$$

and (3.1), we get the

Procedure:

$$U_n^0 = P,$$

$$\alpha_{n+1}^k = U_{n+1}^{k-1}(z_n^k)/U_n^k(z_n^k),$$

$$U_{n+1}^k = (U_{n+1}^{k-1} - \alpha_{n+1}^k \, U_n^k)/(z - z_n^k),$$

$$z_{n+1}^k = -\alpha_{n+1}^k.$$

This method has been proposed in this form by Bauer and Samelson [3]. For $p = 1$, this is a variation of Bernoulli's method.

3.4 A generalized method of Samelson

The procedure considered in the last section can be generalized by choosing the m_k arbitrarily. In this way, one obtains the following

Procedure:

$$(3.2) \begin{cases} U_n^0 = P, \\ w_{n+1}^k U_n^k + u_n^k U_{n+1}^k = U_{n+1}^{k-1}, \\ u_{n+1}^k = u_n^k + w_{n+1}^k. \end{cases}$$

This is exactly Newton's method for the equation (2.3), i.e. the procedure (2.4) with $x_n^k = u_n^k$, and $X_n^k = U_n^k$.

K. Samelson [10] constructed this method without using an operator like G. He considered the ordinary Newton method for calculating a root of a polynomial as gained from a "definite" method by truncation, and shifts of the origin. By properly defining a "shift of the origin for a definite method of factorization" (Nullpunktverschiebung bei einem definiten Faktorisierungsverfahren) he derived some iterative methods which include the procedure mentioned above.

3.5 The ordinary Newton method

Newton's method

$$(3.3) \qquad z_{n+1} = z_n - P(z_n)/P'(z_n)$$

for calculating a root of $P(z)$ can also be considered as a special case of the general procedure (2.3), (2.4).

Choose $p = 1$, $m = 1$,

$$x_n = u_n, \quad X_n = (P - P(z_n))/(z - z_n),$$

and write

$$w_{n+1} = \alpha_{n+1}, \quad u_n = z - z_n$$

with constants α_{n+1}. Then, one obtains from (2.4) the

Procedure:

$$U_n^0 = P,$$

$$X_n = (P - P(z_n))/(z - z_n),$$

$$\alpha_{n+1} = P(z_n)/X_n(z_n),$$

(3.4) $\quad z_{n+1} = z_n - \alpha_{n+1},$

$$U_{n+1} = (P - \alpha_{n+1} X_n)/(z - z_n).$$

Because $X_n(z_n) = P'(z_n)$, formula (3.4) is equivalent to (3.3). Note that the numbers α_{n+1} can be calculated without computing the polynomials U_n.

3.6 Bairstow's method

For a given polynomial $u(z)$ of exact order $m (1 \leq m \leq N-1)$ and leading coefficient 1, the polynomial $P(z)$ may be written as

$$P(z) = u(z) U(z) + v(z)$$

where $U(z)$ is a polynomial of order $N - m$ and $v(z)$ is a polynomial of order $m - 1$. We want to choose $u(z)$ such that $v(z) \equiv 0$.

If $u(z)$ and $v(z)$ are given in power representation:

(3.5) $\quad u(z) = z^m + \gamma_{m-1} z^{m-1} + \ldots + \gamma_1 z + \gamma_0$,

(3.6) $\quad v(z) = \delta_{m-1} z^{m-1} + \ldots + \delta_0$,

the coefficients δ_i are functions of the coefficients γ_j. The problem is the to solve the equations

$$\delta_i(\gamma_0, \ldots, \gamma_{m-1}) = 0 \quad (i = 1, 2, \ldots, m-1).$$

If one applies Newton's method to these equations, one obtains a procedure for splitting the polynomial $P(z)$ into two factors u and U.

For $m = 1$, this procedure is the ordinary Newton method (3.3). For $m = 2$, this describes the method of Bairstow (see, for example, [5], [7]).

For all m ($1 \leq m \leq N-1$) this procedure may be written as a special case of (2.4), namely the following

Procedure:

(3.7) $\begin{cases} X_n \text{ is a polynomial of order } N - m, \text{ such that} \\ P = u_n X_n + v_n \\ \text{with a polynomial } v_n \text{ of order } m-1, \\ w_{n+1} X_n + u_n U_{n+1} = P, \\ u_{n+1} = u_n + w_{n+1}. \end{cases}$

Notice, that again the polynomials u_n may be gained without constructing the polynomials U_n.

4. Convergence and Estimation

One advantage in writing different methods as special cases of a general procedure like (2.3) is the possibility of developing a unified theory of convergence and error estimation. For the procedure (2.3), this has not yet been worked out. We give here only a few general remarks.

For a given polynomial $v(z)$ of exact order q, a norm $||v||$ may be defined in many different ways. For example, v may be considered as a function defined on some domain, and one of the usual function norms may be used. On the other hand, $v(z)$ can be represented by an $(m+1)$-dimensional vector $\hat{v}$, and one may use $||v|| = ||\hat{v}||$ with some vector norm $||\hat{v}||$.

Whatever norms are chosen for the components u^k, U^k of the elements $\mathbf{u} \in R$, and the components s^k of $\mathbf{s} \in S$, by defining

$$\nu[\mathbf{u}] = (||u^1||, \ldots, ||u^p||, ||U^1||, \ldots, ||U^p||)^T,$$

$$\mu[\mathbf{s}] = (||s^1||, \ldots, ||s^p||)^T,$$

one obtains vector valued norms as they can be used in the general convergence theory described in [11].

From these generalized norms, one may obtain real valued norms

(4.1) $\qquad ||\mathbf{u}|| = ||\nu[\mathbf{u}]||, \quad ||\mathbf{s}|| = ||\mu[\mathbf{s}]||$

by using some vector norms $||v||, ||\mu||$.

The application of the theory in [11], or any other convergence theory for Newton's method, to the operator (1.1) is particularly simple in the following sense. The second derivative $G''(\mathbf{x})$ of the operator G is constant. That means, it does not depend on $\mathbf{x}$, and therefore, one can use global bounds of this derivative which do not depend on the domain of the elements $\mathbf{x}$ in consideration. Such bounds of G'' occur, usually in one way or another, in the convergence theories for Newton's method.

To illustrate this fact, we consider the special case $p = 1$ and assume that the equation $G\mathbf{u} = \mathbf{0}$ has a solution $\mathbf{u}^*$. One derives the following formula:

$$G'(\mathbf{x}_n)(\mathbf{u}^* - \mathbf{u}_{n+1})$$

$$= [G'(\mathbf{x}_n) - G'(\mathbf{u}_n)](\mathbf{u}^* - \mathbf{u}_n) - G\mathbf{u}^* + G\mathbf{u}_n + G'(\mathbf{u}_n)(\mathbf{u}^* - \mathbf{u}_n)$$

$$= -(x_n - u_n)(U^* - U_n) - (u^* - u_n)(X_n - U_n) - (u^* - u_n)(U^* - U_n) .$$

If the norms of the occuring polynomials are chosen such that

$$||u \cdot U|| \leq \alpha ||u|| \, ||U||$$

with a parameter α, possibly depending on the degrees m and $N-m$, one obtains the estimate

$$\|G'(\mathbf{x}_n)(\mathbf{u}^* - \mathbf{u}_{n+1})\| \leq$$

$$\alpha \cdot \begin{pmatrix} \|x_n - u_n\| \\ \|X_n - U_n\| \end{pmatrix}^T \begin{pmatrix} 0 & 1 \\ 1 & 0 \end{pmatrix} \begin{pmatrix} \|u^* - u_n\| \\ \|U^* - U_n\| \end{pmatrix}$$

$$+ \frac{\alpha}{2} \begin{pmatrix} \|u^* - u_n\| \\ \|U^* - U_n\| \end{pmatrix}^T \begin{pmatrix} 0 & 1 \\ 1 & 0 \end{pmatrix} \begin{pmatrix} \|u^* - u_n\| \\ \|U^* - U_n\| \end{pmatrix}$$

(4.2) $\quad = \alpha \cdot (\nu[\mathbf{x}_n - \mathbf{u}_n] + \frac{1}{2}\nu[\mathbf{u}^* - \mathbf{u}_n], A\nu[\mathbf{u}^* - \mathbf{u}_n])$

with the notations

$$\nu[\mathbf{u}] = \nu[(u,U)] = \begin{pmatrix} \|u\| \\ \|U\| \end{pmatrix}, \quad A = \begin{pmatrix} 0 & 1 \\ 1 & 0 \end{pmatrix}$$

and the usual inner product (,) for vectors.

The right hand side of the inequality above is of a very simple form. The occuring matrix A is constant because $G''(\mathbf{x})$ is constant.

With (4.1) and an appropriate matrix norm, one derives from (4.2)

(4.3) $\quad \|G'(\mathbf{x}_n)(\mathbf{u}^* - \mathbf{u}_{n+1})\| \leq \alpha \|A\| (\|\mathbf{x}_n - \mathbf{u}_n\| + \frac{1}{2}\|\mathbf{u}^* - \mathbf{u}_n\|) \|\mathbf{u}^* - \mathbf{u}_n\|.$

Of course, in order to get an estimation for $\mathbf{u}^* - \mathbf{u}_{n+1}$, a bound for the operator $[G'(\mathbf{x}_n)]^{-1}$ is needed. In a neighborhood of some fixed element $\mathbf{w}$, say $\mathbf{w} = \mathbf{u}_0$, or $\mathbf{w} = \mathbf{u}^*$, this bound may be obtained from

(4.4) $\qquad [G'(\mathbf{x}_n)]^{-1} = [G'(\mathbf{w})]^{-1} [I - G'' \bullet (\mathbf{w}-\mathbf{x}_n)]^{-1}$

by again using the fact that G'' is constant.

From these relations (4.3), (4.4), the quadratic convergence of Newton's method $(\mathbf{x}_n = \mathbf{u}_n)$ is easily deduced in case $[G'(\mathbf{u}^*)]^{-1}$ exists on S.

One may also consider the possibility of obtaining theorems on monotonic convergence by using general results of Kalaba [8], and Collatz [4]. Some remarks on monotonic behavior of approximations in factorization methods have been made by Samelson [10].

5. Other applications of Newton's method. Comparison

There are many ways of defining an operator G such that solving the equation $G\mathbf{u} = \mathbf{0}$ is equivalent to splitting the polynomial $P(z)$ into factors. Applying Newton's method to any such equation yields an approximation procedure for factorization. We describe some possible ways of proceeding. In order to explain the ideas, we can restrict ourselves to the case $p = 1$.

5.1 Formulating the problem of factorization

Let u be a given polynomial of degree m, with leading coefficient 1, and let β denote a fixed integer with $m-1 \leq \beta \leq N-1$.

Then, there are polynomials U of degree $N - m$ with leading coefficient c such that $s = P - u\,U$ is a polynomial of degree at most β.[*)]

For $\beta = N - 1$, all polynomials U of degree $N - m$ with leading coefficients c have this property. For $\beta = m - 1$, there is exactly one such polynomial U. This polynomial U can be considered as the image of u under an operator B: $U = Bu$.

For $m - 1 < \beta < N - 1$, the polynomials U depend on u, but are not uniquely determined by u. One may write $U = B\mathbf{u}$ where $\mathbf{u} = (u,V)$ consists of two parts, namely the given polynomial u and a quantity V which describes in some way or other the "free parameters" contained in U for a given u.

For $\beta = N - 1$, one may use $V = U$, $\mathbf{u} = (u,U)$. For $\beta = m-1$, we can write $\mathbf{u} = u$, or $\mathbf{u} = (u,0)$, for consistency.

After having defined V properly for a given β one may define an operator G by

$$G\mathbf{u} = P - u\,B(u,V).$$

on the set of all elements $\mathbf{u} = (u,V)$ of the prescribed form. Then, $G\mathbf{u} = 0$ is equivalent to $P = uU$, and the Newton procedure applied to $G\mathbf{u} = 0$ furnishes an approximation method for splitting P.

5.2 The two limit cases

In order to compare different choices of the integer β in Section 5.1, let us only consider the limit cases $\beta = N-1$, and

---
[*)] The restriction $\beta \leq N-1$ and the restrictions posed on the polynomial U are not necessary. Omitting these restrictions, one obtains still further possibilities of formulating the problem.

and $\beta = m-1$.

The choice $\beta = N-1$ leads to the definition of G in Section 1 (with $m = 1$):

(5.1) $$\mathbf{u} = (u,U) , \quad G\mathbf{u} = P - uU.$$

For $\beta = m-1$, we get

(5.2) $$\mathbf{u} = u , \quad G\mathbf{u} = P - u(Bu) .$$

Newton's method for (5.1) is given by (3.2) with $p = 1$:

(5.3) $$\begin{cases} w_{n+1} U_n + u_n U_{n+1} = P, \\ u_{n+1} = u_n + w_{n+1} . \end{cases}$$

Newton's method for (5.2) is the procedure (3.7) in Section 3.6, except that the occuring U_n are not calculated.

The operator in (5.1) has the important advantage of being very simple in the following sense: $G\mathbf{u}$ is a quadratic expression in $\mathbf{u}$, the second derivative G'' is constant. The significance of this fact for the convergence and estimation theory has already been pointed out in Section 4. It remains to be investigated how this fact influences the numerical behavior of the procedure.

The expression $G\mathbf{u}$ in (5.2) depends in a more complicated way on $\mathbf{u}$. For $m = 1$, we have, for example, $u = z - \alpha$, and $Bu = (P(z) - P(\alpha))/(z - \alpha)$, so that

$$Gu = P(\alpha).$$

The corresponding Newton procedure is the ordinary Newton method (3.3). In the convergence theory of this method, one faces the additional difficulty that bounds on the second derivative depend on the domain considered.

The procedures of Sections 3.5 and 3.6 have now been described in two ways. They are Newton procedures for the operator (5.2), and on the other hand, they are generalized Newton procedures (with $x_n \neq u_n$) for the operator (5.1).

Of course, the difficulties in the convergence theory caused by the more complicated nature of the operator in (5.2) have not disappeared by describing the procedures in the second way using the operator (5.1). These difficulties are now postponed to the estimation of $x_n - u_n$.

But the second description may serve in comparing these procedures with the corresponding Newton procedures (5.3) for the operator (5.1). For example, it is immediately clear that there are strong interactions between the approximations u_n and U_ν of the procedure (5.3), while in the procedure (3.7), the U_ν and X_ν have no influence on the following approximations u_n.

5.3 Comparison of the operations

For the procedures (3.7) and (5.3), we indicate now the "usual way" of proceeding numerically and compare the necessary operations.

In case of the procedure (5.3), the main problem in each step is to solve the equation

(5.4) $$wU + uV = P$$

for $w = w_{n+1}$ and $V = U_{n+1}$ if $u = u_n$ and $U = U_n$ are given. One may proceed as follows using

$$w = \gamma_{m-1} z^{m-1} + \ldots + \gamma_0.$$

Divide the polynomials P, $z^{m-1}U, \ldots, zU, U$ by u:

(5.5) $\quad P = uQ + q$,

(5.6) $\quad z^i U = uY^{(i)} + y^{(i)} \quad (i = m-1, \ldots, 0)$,

with polynomials q, $y^{(i)}$ of degree at most $m-1$, and polynomials Q, $Y^{(i)}$.

Then, (5.4) is equivalent to

(5.7) $\quad \gamma_{m-1} y^{(m-2)} + \ldots + \gamma_0 y^{(0)} = q$,

(5.8) $\quad Q - \gamma_{m-1} Y^{(m-1)} - \ldots - \gamma_0 Y^{(0)} = V$.

The first of these two relations represents a linear algebraic system for the unknown coefficients γ_i of w. The second relation yields V.

In the procedure (3.7), divisions like (5.5), (5.6) have also to be carried out. With the notations already used and $X_n = Q$, $v_n = q$, the equations (3.7) lead to

(5.9) $\quad\quad\quad\quad\quad P = uQ + q$,

(5.10) $\quad z^i Q = u Y^{(i)} + y^{(i)} \quad (i = m-1,\ldots,0)$,

and the system (5.7). The difference is, in (5.10) $Q = X_n$ is used instead of $U = U_n$, and the calculation of $V = U_{n+1}$ is omitted.

The divisions (5.5), (5.6) together require two Horner-m-steps. So do the divisions (5.9), (5.10). The operations in calculating V by (5.8) amount to another Horner-m-step. Thus, in each iteration step, the procedure (3.7) requires fewer operations than the procedure (5.3).

5.4 Numerical Example

In order to illustrate the difference between the method of Bairstow and the Newton procedure (5.3) we consider a very simple example, namely the polynomial

$$P(z) = z^4 - 10z^3 + 37z^2 - 72z + 80.$$

It has a double root at $z = 4$, and a pair of conjugate complex roots $z = 1 \pm 2i$.

As initial approximations for the quadratic factor u, we have used in both methods

$$u_0 = z^2 - 2az + a^2$$

with several values of a. For the Newton procedure (5.3) we need also an initial approximation U_0 for the second factor U. This approximation U_0 has been chosen in two ways:

1. U_0 is the polynomial X_0 of Bairstow's method (see (3.7)),

2. $U_0 = z^2 - 2bz + b^2$ with several values of b.

Table 1 shows the number of iterations required to obtain a certain accuracy[*] using Bairstow's method or procedure (5.3) with $U_0 = X_0$. It is indicated in parentheses to what factor the approximations u_n converge. (d) indicates convergence to $(z - 4)^2$, (c) convergence to $(z - 1)^2 + 4$.

For some values of a, relatively many iteration steps are necessary. In Newton's procedure (5.3), it is possible to reduce the number of these steps by choosing an approximation $U_0 = (z - b)^2$. Table 2 shows how many steps are necessary for $a = 2.2$, and various values of b.

This is only a very simple example. But it hints at the obvious fact that the advantage of Bairstow's method of requiring fewer operations in each iteration step is not the only point to be considered. In this connection notice also that Bairstow's method is related to procedure (5.3) in approximately the same manner as the ordinary Newton method (3.3) is related to the method in Section 3.3 with $p = 1$.

---

[*] The computation was stopped when two successive approximations differed in both coefficients by less than 0.00002. One gets almost the same numbers by stopping the procedure when the exact errors of those coefficients are smaller than certain prescribed bounds.

Table 1: Number of iterations for B = Bairstow's method with
$u_0 = (z - a)^2$, and N = Newton's method (5.3) with
$u_0 = (z - a)^2$, $U_0 = X_0$.

a	B	N
-15.0	12 (c)	11 (c)
-10.0	11	10
- 8.0	10	10
- 5.0	9	8
- 3.0	8	7
- 1.0	7	6
- 0.5	6	6
0.5	6	5
1.0	6	6
2.0	12	9
2.2	28	22
2.4	14 (c)	9
2.5	14 (d)	9 (c)
2.6	10	10 (d)
2.7	14	10
2.8	25	15
3.0	11	9
3.2	8	8
3.5	6	5
4.5	5	5
5.0	7	6
6.0	8	6
8.0	9	8
10.0	11	9
15.0	11	10
20.0	12 (d)	11 (d)

T a b l e 2: Number of iterations for N^* = Newton's method (5.3) with $u_0 = (z - 2.2)^2$, and $U_0 = (z - b)^2$.

b	N^*
3.0	8 (c)
3.2	7
3.5	6
4.5	5
5.0	6
6.0	8
8.0	8
10.0	9
15.0	8
20.0	10 (ċ)

5.5 Implicit definition of the operator

As a simple example for still further possibilities of defining an appropriate operator G, we consider the case where $\beta = m-1$, $m = 2$ in Section 5.1. We define Bu as in that section. However, we do not use definition (5.2).

Instead, we write

$$P - u(Bu) = C(u, Gu)$$

where C is a polynomial depending on u and Gu in the following way:

$$C(u, Gu) = a(z - \gamma) + b$$

for $u = z^2 - \gamma z - \delta$, $Gu = az + b$.

The Newton procedure for this operator G is the version of Bairstow's method which is described in the book of P. Henrici [6].

REFERENCES

1. F.L. Bauer: Direkte Faktorisierung von Polynomen. Sitz.Ber. Bayer. Akad.Wiss. 1956, 163-309.

2. F.L. Bauer: Das Verfahren der Treppeniteration und verwandte Verfahren zur Lösung algebraischer Eigenwertprobleme. Z.Angew. Math.Phys. *8*, (1957) 214-235.

3. F.L. Bauer und K. Samelson: Polynomkerne und Iterationsverfahren. Math.Z. *67*, (1957) 93-98.

4. L. Collatz: Monotonie und Extremalprinzipien beim Newtonschen Verfahren. Numer.Math. *3*, (1961) 99-106.

5. R.W. Hamming: Numerical Methods for Scientists and Engineers, 1962, 356 ff.

6. P. Henrici: Elements of Numerical Analysis, 1964, 110 ff.

7. E. Isaacson and H.B. Keller: Analysis of Numerical Methods, 1966, 131 ff.

8. R. Kalaba: On nonlinear Differential Equations, the Maximum Operations and monotone Convergence. J. Math. Mech. *8*, (1959) 519-574.

9. H. Rutishauser: Der Quotienten-Differenzen-Algorithmus. Mitt. Inst.angew.Math. ETH Zürich Nr. 7. Basel-Stuttgart 1957.

10. K. Samelson: Faktorisierung von Polynomen durch funktionale Iteration. Bayer.Akad.Wiss., Math.-Nat. Klasse, Abhandlgn., Neue Folge, Heft 95, München 1959.

11. J. Schröder: Ueber das Newtonsche Verfahren. Arch.Rat.Mech. Anal. *1*, (1957) 154-180.

Prof. J. Schröder
Universität zu Köln
D-5 Köln-Lindenthal
Germany

E. Specker

The Fundamental Theorem of Algebra in Recursive Analysis

N is the set of natural numbers. Q is the field of complex rational numbers, i.e. the field of numbers $a+bi$, where a and b are rational. A sequence σ of elements of Q (i.e. a function $\sigma, \sigma: N \to Q$) is recursive iff there exist recursive functions f_j ($j=1,2,3,4$; $f_j: N \to N$) such that for all n in N

$$\sigma(n) = \frac{f_1(n)}{f_2(n)+1} + \frac{f_3(n)}{f_4(n)+1} i \, .$$

Such a sequence σ is recursively convergent iff there exists a recursive function k ($k: N \to N$) such that for all h, j, n in N

(1) $\quad h \geq k(n) \wedge j \geq k(n) \Rightarrow |\sigma(h) - \sigma(j)| < \dfrac{1}{n+1}$.

A complex number c is recursive iff there exists a recursive sequence σ ($\sigma: N \to Q$) which converges recursively and such that

$$c = \lim_{n \to \infty} \sigma(n) \, .$$

If k is a function satisfying (1) for all h, j, n in N, then for all h, n in N

321

(2) $$h \geq k(n) \Rightarrow |\sigma(h) - c| \leq \frac{1}{n+1}.$$

(For these definitions, see [2]).

It is not difficult to show that the recursive complex numbers form a field (which is an extension of **Q**); furthermore, the field of recursive complex numbers is algebraically closed, i.e. the roots of a polynomial with recursive complex coefficients are recursive complex numbers (cf. [4]).

This theorem might be called the "weak fundamental theorem of algebra in Recursive Analysis". Its weakness is obvious from the following outline of a proof: For every polynomial p there exists a polynomial q such that (1) all the roots of p are roots of q, (2) q has no multiple roots, (3) if the coefficients of p are recursive complex numbers, so are the coefficients of q. In order to prove the weak fundamental theorem, it suffices therefore to prove it for polynomials without multiple roots; this is done easily (and effectively) on the basis of known proofs of the classical fundamental theorem. The non-effectiveness of the above proof lies in the passage from the polynomial p to the polynomial q; this passage cannot be made effective and the weak fundamental theorem does not answer the following question: Let σ_i, k_i be sequences $(0 \leq i \leq m;\ \sigma_i: N \to Q;\ k_i: N \to N)$ such that for all h, j, n in N and all i, $0 \leq i \leq m$, $h \geq k_i(n) \wedge j \geq k_i(n) \Rightarrow |\sigma_i(h) - \sigma_i(j)| < \frac{1}{n+1}$; assume furthermore (and provisionally) $\sigma_m(n) = 1$ for all n. Does there exist an effective procedure for transforming the sequences σ_i, k_i $(0 \leq i \leq m)$ into sequences τ, ℓ $(\tau: N \to Q;\ \ell: N \to N)$ such that for all h, j, n in N

$$h \geq \ell(n) \wedge j \geq \ell(n) \Rightarrow |\tau(h) - \tau(j)| < \frac{1}{n+1}$$

and

$$\lim_{n \to \infty} \sum_{k=0}^{m} a_k(n) (\tau(n))^k = 0$$

(i.e. putting $s_k = \lim_n a_k(n)$, $0 \leq k \leq m$, $t = \lim_n \tau(n)$, we have $\sum_0^m s_k t^k = 0$)?

In order to prove that such a procedure in fact exists, we study the map associating the system of roots to a polynomial. Let therefore m be a fixed positive integer. Two polynomials $\sum_{k=0}^{m} s_k x^k$ and $\sum_{k=0}^{m} s'_k x^k$ have the same roots iff $(s_0, \ldots, s_m)$ and $(s'_0, \ldots, s'_m)$ are proportional, i.e. if they represent the same point in projective m-space. Let P^m be the complex projective m-space and let P^m_r be the subset of those points in P^m which can be represented by a complex rational $(m+1)$-tuple. P^m_r is dense in P^m. We introduce a metric on P^m (compatible with its topology) satisfying the following conditions: The distance function has rational values on $P^m_r \times P^m_r$ and is recursive on $P^m_r \times P^m_r$. The distance between two points u, v is denoted by $|u-v|$.

The space of root-systems of polynomials of degree at most m is the symmetric product T^m of m two-spheres, S^2. Indeed, S^2 has to be chosen, and not the complex plane, because we have not excluded polynomials with leading coefficient zero and have therefore to admit roots equal to infinity. A point in T^m is represented by an m-tuple $(p_1, \ldots, p_m)$ (p_i, $1 \leq i \leq m$, being points of S^2); two m-tuples $(p_1, \ldots, p_m)$ and $(p'_1, \ldots, p'_m)$ represent the same point of

T^m iff there exists a permutation π such that for all i, $1 \le i \le m$,

$$p'_i = p_{\pi(i)} .$$

Let T^m_r be the subset of T^m representable by m-tuples $(p_1,\ldots,p_m)$ such that all the p_i, $1 \le i \le m$, are complex rational numbers. (It is left to the reader to consider infinity as rational or irrational.) T^m_r is dense in T^m.

We introduce a metric on T^m (compatible with its topology, and also denoted by $|u-v|$) such that the following conditions are satisfied: The distance function has rational values on $T^m_r \times T^m_r$ and is recursive on $T^m_r \times T^m_r$. There exists a recursive function ν on N whose values are k-tuples of points of T^m_r such that $\nu(n)$ is a $\frac{1}{n}$-net of T^m, i.e. $\nu(n)$ is a k-tuple (k depending on n) $(q_1,\ldots,q_k)$ of points of T^m_r such that for every point q of T^m there exists an index j, $1 \le j \le k$, such that

$$|q - q_j| < \frac{1}{n} .$$

There exists a natural map f from T^m into P^m, viz. the map associating the polynomial to a system of roots. f maps T^m_r recursively into P^m_r and is recursively continuous, i.e. there exists a recursive function κ (from positive rationals to positive rationals) such that for all u,v in T^m and all $\delta, \delta > 0$,

$$|u-v| < \kappa(\delta) \Rightarrow |f(u) - f(v)| < \delta .$$

The classical fundamental theorem of Algebra states that f is a map from T^m onto P^m. The map f being one-to-one and the space T^m being compact, it follows that the inverse function f^{-1} is continuous and that the spaces T^m, P^m are homeomorphic (cf. [1], p. 95 and [3], p. 392). The strong fundamental theorem of algebra in Recursive Analysis states that f^{-1} is a recursively continuous and recursive topological function on P^m relative to P^m_r. (The notion of a recursive topological function is the straightforward generalization of recursive real function [2]; the present situation is so simple that it is not necessary to introduce the notion explicitly.) The recursiveness of f^{-1} is an immediate consequence of the three following assertions: (1) f^{-1} is recursive on the image $f[T^m_r]$ of T^m_r by f. (2) The set $f[T^m_r]$ is a recursive subset of P^m_r. (3) $f[T^m_r]$ is recursively dense in P^m_r, i.e. there exists a recursive function $\Delta, \Delta : N \times P^m_r \to f[T^m_r]$, such that for all n in N and all p in P^m_r

$$|p - \Delta(n,p)| < \frac{1}{n+1}.$$

The recursive continuity of f^{-1} is an immediate consequence of the following

<u>Theorem</u>
 Let f be a function mapping T^m onto P^m and satisfying the following conditions:

(1) The restriction of f to T^m_r is recursive and maps T^m_r into P^m_r.

(2) f is recursively continuous, i.e. there exists a recursive function κ (from positive rational numbers to positive rational numbers) such that for all u, v of T^m and all $\delta, \delta > 0$,

$$|u - v| < \kappa(\delta) \Rightarrow |f(u) - f(v)| < \delta.$$

Then there exists a recursive function λ (from positive rational numbers to positive rational numbers) such that for all u,v of T^m and all δ, $\delta > 0$,

$$|u - v| \geq \delta \Rightarrow |f(u) - f(v)| \geq \lambda(\delta) .$$

P r o o f . Let δ be rational positive and ε such that $1/\varepsilon$ is a natural number and $\varepsilon < \delta/4$. Let $(x_1,\ldots,x_k)$ be the ε-net $\nu(1/\varepsilon)$ and define the rational number η as follows:

$$\eta = \min_{\substack{1 \leq i < j \leq k \\ |x_i - x_j| \geq \frac{\delta}{2}}} |f(x_i) - f(x_j)|.$$

(If there are no i,j such that $|x_i - x_j| \geq \delta/2$, put $\eta = \infty$.)

Given the net, the number η can be determined effectively, i.e. η is a recursive function of $1/\varepsilon$. There exists a positive number η_0 such that $\eta_0 \leq \eta$ for all nets satisfying the condition imposed on ε. The existence of such an η_0 follows (non-constructively!) from the continuity of f^{-1}: Given $\delta/2$, $\delta/2 > 0$, there exists η_0, $\eta_0 > 0$, such that for all u,v of T^m

$$|u - v| \geq \frac{\delta}{2} \Rightarrow |f(u) - f(v)| \geq \frac{\eta_0}{2} .$$

It is convenient for the following to assume that the function κ is monotonically increasing; if κ does not have this property it is not difficult to define κ' having this additional property.

An ε-net is admissible if $\varepsilon < \kappa(\eta/3)$. The following assertions are easily verified:

(1) It is decidable whether a given ε-net is admissible.

(2) There exist admissible ε-nets. (If $\varepsilon < \kappa(\eta_0/3)$ then the net is certainly admissible; this is the point where we make use of the monotonicity of κ.)

Let η_1 be the number η corresponding to the first ε-net in the enumeration ν; η_1 depends recursively on δ.

We shall show that for all u,v in T^m

$$|u - v| \geq \delta \Rightarrow |f(u) - f(v)| \geq \frac{\eta_1}{3}.$$

Having shown this and defining $\lambda(\delta)$ to be $\eta_1/3$, the proof of the theorem will be completed.

Assume that u,v are points of T^m such that $|u-v| \geq \delta$ and let $(y_1,\ldots,y_n)$ be the first admissible ε-net, $0 < \varepsilon \leq \delta/4$. There exist indices i, j $(1 \leq i,j \leq n)$ such that

$$|u - y_i| < \varepsilon \leq \frac{\delta}{4}$$

$$|v - y_j| < \varepsilon \leq \frac{\delta}{4}.$$

Because of $|u-v| \geq \delta$, we have

$$|y_i - y_j| \geq \frac{\delta}{2}$$

and therefore

$$|f(y_i) - f(y_j)| \geq \eta_1.$$

Because of

$$|u - y_i| < \varepsilon \leq \kappa(\frac{n_1}{3}) \quad \text{and}$$

$$|v - y_j| < \varepsilon \leq \kappa(\frac{n_1}{3})$$

we have

$$|f(u) - f(y_i)| < \frac{n_1}{3} \quad \text{and}$$

$$|f(v) - f(y_j)| < \frac{n_1}{3}$$

and therefore

$$|f(u) - f(v)| \geq \frac{n_1}{3} = \lambda(\delta) \; .$$

R e m a r k . The above theorem on the recursive continuity of f^{-1} can be extended to general recursive metric spaces which are totally bounded. This generalization is the constructive analogue of the theorem asserting the continuity of the inverse function of a continuous one-to-one function defined on a compact space. The proof itself, however, is not constructive. Constructive proofs establishing the strong fundamental theorem of algebra in Recursive Analysis are well known, e.g. [5].

REFERENCES

1. P. Alexandroff und H. Hopf: Topologie, Berlin, Springer, 1935.

2. R.L. Goodstein: Recursive Analysis, Amsterdam, North-Holland Publishing Company, 1961.

3. P.J. Hilton and S. Wylie: Homolpgy Theory, Cambridge, University Press, 1960.

4. H.G. Rice: Recursive real numbers. Proc.Amer.Math.Soc. 5 (1954) 784 - 791.

5. H. Weyl: Randbemerkungen zu Hauptproblemen der Mathematik, Math. Zeitschrift *20* (1924), 131-150.

Prof. E. Specker
Eidgenössische Technische Hochschule
CH-8006 Zürich
Switzerland

Index

A

Abbrechkriterium (*see also* stopping criterion), 285
absolute free algebra, 125,126,127
acceleration (*see also* Steffensen accelerating procedure), 216
acceptance criterion, 43
accuracy (*see also* precision), 13,58,169
Adams, D., 167
Aitken, A.C., 186
algebraically closed, 322
a posteriori
  estimate, 174
  upper bound, 103
argument principle, 83
arithmetic (*see also* numbers),
  exact, 42
  floating-point, 53
arithmetization, 117
Asser, G., 127
asymptotic error constant, 159
Auslöschung (*see also* cancellation), 190,283,284,292
Axt, P., 123

B

Bairstow's method (*see also* Newton-Bairstow procedure), 295,305,306,315,316,317,319

Bareiss, E.H., 135
Bauer, F.L., 181,182,184,285, 290,291,296,303
Bernoulli, D., 181,182
Bernoulli method, 181,184,189, 191,295,302,303
  Lagrange form of, 302
Bernoulli-Lagrange'sche Methode, 288
Besselfunktion, 287
bigradient, 131,132,133,138,139, 148,149
  polynomial, 133,138
biorthogonal system, 186
bisection, 37,45,50,75
Bôcher, M., 135
Boolean variables, 57
Borel, E., 138, 150
Brodetzky-Smeal, Verfahren von, 289

C

cancellation (*see also* Auslöschung), 187,188,190
Cauchy's existence proof, 209
Cauchy index, 148
Cayley-Parametrisierung, 291
characteristic function, 121
charakteristisches Polynom, 287
Church, A., 118,119,121,123
circle method, 207,208

circumscribed rectangles, 85
Clenshaw, C.W., 284
Collatz, L., 310
common divisor, 134
common factor, highest, 70,72, 74
companion matrix, 182
computable, 118
   function, 115,116,117,119,120
   real number, 116
   sequence, 116,119
computably convergent, 116,119
computer, ideal, 53
constructivity, 115
constructive form, 69
constructive solution, 115
continued fraction, 131,141,146, 147,148
contrapoint, 41
convergence, 1,2,21,37,213,307
   conditions of, 230,231
   global, 154,160
   linear, 2,39,40,80,93,232, 243,246,248,250,252
   linear, uniform, 77
   monotonic, 310
   numerical, 241
   order of, 49,235,247,257, 263,266
   quadratic, 2,159,168,198,246, 252,257,310
   rate of, 158,207,208,215
   recursive, 75
   slow, 42
   sufficient condition of, 245
   superlinear, 40,234,246
   unified theory of, 295,307
convergence function, 82
convergence theorem, abstract, 1,14,17
convex function, 42

D

decidable, 119,327
deflation, 65,166,172,175
   backward, 153
   stable, 151,153
Dekker, T.J., 49
Derwidué, L., 148
digits, guarding, 66
   significant (*see also* figures, significant), 53
Dijkstra, E.W., 48
dominance, numerical, 187,188
down hill method (*see also* steepest descent), 197

E

effectiveness, 322
effective operations, 69,71
elementary inductive definition, 118,121,123,124,125,127
elementary inductive definition, generalized, 127
   restricted, 122
eliminant, 70
equation, cubic, 174
   quadratic, 53,174,226
   quartic, 174
error (*see also* round-off error), 8,159,211
   absolute, 12
   analysis, backward, 64
   bound (*see also* a posteriori), 1,2,103
   estimation, unified theory of, 307
   induced, 66
   relative, 12,13
η-inclusion set (*see also* inclusion set), 78,79
Euclidean algorithm, 138,148
exclusion algorithm, 100

exclusion radius, 96
exclusion test, 79,80,81,86
  one-sided, 88,94,95,96
  one-sided, uniformly convergent, 87,94,95
  uniformly convergent, 99
extrapolation, 257
  linear, 37,48
  successive linear, 39

F

factorization, 296
Farkas, I., 167
Fehlerschranken-Arithmetik (see also interval arithmetic), 1,2,8
Fibonacci numbers, 216
figures, significant (see also digits, significant), 170
Filterproblem, 290
fixed point, 15,230,231,236, 237,261
Forsythe, G.E., 175
Fréchet derivative, 232,298
Fréchet differential, 269
Frobenius, G., 131,143
function, computable, 115,116, 117,119,120
  elementary, 122
  γ-primitive recursive, 125
  genuine primitive recursive, 125,126
  μ-recursive, 117,118
  ordinary primitive recursive, 125,126,127
  primitive recursive, 121,123, 124,127
  recursive, 118,119,120,121, 123,124,321
  regular, 118,119,120
  Turing-computable, 118

fundamental theorem of algebra, 77,115
  strong, 325,328
  weak, 322

G

Gantmacher, F.R., 149,226
Gauss, C.F., 87
Gauss proof, second, 69
Gödel, K., 118
Gödelization, 117,125,126,127
Goodstein, R.L., 117,119
Graeffe's method, 169,288,289, 290
Grzegorczyk, A., 121

H

Hankel determinants, 135,138, 144,145,148
Heinermann, W., 123
Henrici, P. 21,152,172,319
Hermes, H., 118,120
hierarchy, 115,121,123
Hölder norm, 235,244,248
Horner-m-step, 315
Horner scheme, 95,96,102,284
Horner unit, 214
Hurwitz, A., 205

I

IBSYS operating system, 59
ill-conditioning, 65
inclusion algorithm, 79,80
inclusion set (see also η-inclusion set), 81
instability (see also stability), 208
  induced, 64
  inherent, 65

integer exponent, 54
    significand, 54
interactive environment, 153
interpolation, 257
    linear, 37,38,48,50
    successive linear, 39
interval, 37
interval of sign-change, 49
interval reduction, 40
    arithmetic (see also
        Fehlerschranken-
        Arithmetik),1,67
Interval Analysis, 67,221
inverse semi-group, 272
iteration, linear, 251
    order of the, 158
iteration function, 156,158,
    159,174,227,256,265
    general, 255
    matricial, 242,257
    multipoint, 228
    of Newton, 262
    one-point, 227,228
    recursively formed, 260
    scalar, 257

J

Jensen variable, 45
Jordan curves, 83

K

Kahan, W., 53,59,60,80,167,175
Kalaba, R., 310
Keller, H.B., 240,275
Kerner, I.O., 78
Kleene, S.C., 118,120,124
Kleene normal form theorem,
    120
Korganoff, A., 236,241,272,275

L

Laasonen, P., 255
Lagrange'sche Interpolations-
    formel, 284
Laguerre iteration, 174
Laplace expansion, 144
Lehmer, D.H., 99,149

M

Mahn, F.K., 127
Mailloux, B.J., 48
manipulable objects, 116,117,125,
    126
manipulable real number, 116
mapping, effective, 117
Marden, M., 165
method of steepest descent
    (see steepest descent)
monotonic algorithm, 14
Moore, R.E., 67
μ-operator, 118,121
Muir, Th., 135,141
Muller iteration, 173
multilength arithmetic, 65
multiplicity, approximate, 211
Meyer, A.R., 123

N

Netto, E., 150
Newton-Bairstow procedure, 198
Newton direction, 211
    interpolation formula,
        300
    method, 227,247,252,262,267,
        268,284,289,299,304,306,
        310,316,317
    -Modifikation, 288
    procedure, 313,315,319

Newton procedure, generalized
 295,298,299,300,313
 -Raphson iteration, 156,161,
  165,173,174,211,214
 's quotient, 210
Nickel, K., 67,282
non-effectiveness, 322
norm, 193
 generalized, 307
normal form theorem, 124
Nullstellen (*see also* zeros),
 fast mehrfache, 291
 mehrfache, 291
number base, 53
numbers (*see also* arithmetic),
 floating-point, 54
 normalized floating-point,
  53

O

order of convergence (*see*
 convergence)
Orthogonalpolynom, 282,288
Ostrowski, A.M., 65
overflow, 44,48,53,54,55,57,60

P

Padé table, 131,142,143,144,147,
 148
Partialbruchzerlegung, 289
Pavel-Parvu, M., 236,241,272,
 275
perturbed polynomial, 64
 problem, 63
Péter, R., 127
Pol, 287,288,289
 schwacher, 289
 starker, 289
polynomial matrix equation, 226
Post, E., 118
Potter, J.E., 226

power method, 164,169
precision (*see also* accuracy),
 8,182,187,190
 double, 60,168,221
 higher (*see also* multi-
  length arithmetic), 166
 long, 101,102,105
 multiple, 57,59,181
 short, 101,102
 simple, 221
 single, 59,167,220
predecessor-relation, 127
pregnant components, 82
primitive recursivity (*see*
 recursiveness)
projective m-space, 323
pseudo-inverse, 236,238,239,241,
 248,269,272,273,275
purification, 172,175

Q

qd algorithm, 78,131,136,146,147,
 301
 progressive form of the, 138

R

random coefficients, 105,106
Rayleigh approximation, 164
recursive complex number, 321,322
 function, 115
recursiveness, recursivity, 123,
 325
 primitive, 124,125,126
recursively continuous, 324,325
 convergent, 321
 convergent, primitively, 71,
  72,73,74,75,76
regula falsi, 38
Reid, W.T., 226
resolution, relative power of,
 191

resultant, 134
Riccati, matrix equation of, 226
Rödding, D., 128
roots (see zeros)
round-off error (rounding error), 1,2,21,42,44,53 80,103,166,221,241
round-off noise, 199
Routh-Hurwitz criterion, 148
Ruffini-Horner division, 220
Rutishauser, H., 12,78

S

saddle point, 197
Samelson, K., 285,296,303
scaling, 166
Schröder's method, 267
Schur, I., 99,148,202
Schur-Cohn, algorithm, 80
    criterium, 103
Schwarz, H.R., 283
search method, 207
search problem, 193
sequence, primitive, 71,77
    recursive (see also recursive), 321
sign change, 46
simplicity of a solution, 115
Specht, W., 288
Specker, E., 79,124
spectral norm, 249
square method, 207,208
stability (see also instability), 184
steepest descent, method of (see also down hill method), 197,212,214

Steffensen accelerating procedure, 214
Stellenauslöschung (see Auslöschung)
stopping criterion, 7,8,44
Sturm sequence, 141,148
substitution, 117,122,124,125, 126
suspect square, 81,82,83,99, 102,105
Sylvester's method, 70
systems requirements, 60

T

termination (see also stopping criterion), 167,175
    criterion, 151,162,164
test, convergence function of, 82
    convergent, 82,85,86
    effectiveness of a, 86,94,95, 96
    one-sided, 83,85,86,100
tolerance, 43,44,45,207
totally finite set, 128
translation, 151,154,161,162, 163,165,166,168,174,186,187
    complex, 174
    double, 175
Traub, J.F., 65,154,157,159,161,166, 168,227,257,260,267
Treppeniteration, 284,285,288,291, 295,301,302
    abgekürzte, 291
    Newton-type, 303
Treppen-Polynome, 284
Tridiagonalmatrix, 282
Triplex ALGOL, 1
Trudi, N., 131,134,135,141,150
Tschebyscheff-Polynom, 283,284
Turing, A.M., 118

U

uncertainty, 78,80,82,90,93, 100
underflow, 7,21,53,54,55,57, 58,60
unified convergence and estimation theory, 295
uniform convergence, 77,79,90
uniformly linearly convergent, 100
unilateral polynomial equation, 226
unsolvability, 115

V

van Wijngaarden, Ir.A., 48

W

Ward, J.A., 195,196,197
Watkins, B.O., 172
Weierstrasse, K., 78
Weyl, H., 77,79,83
Wheeler, D.J., 50
Wilkinson, J.H., 13,50,64, 65,152,153,169.175
Wronski identities, 136

Z

zero(s) (*see also* Nullstellen),
  counting of, 131
  close, 106
  cluster of, 8,13,181,199, 211,217,221
  equimodular, 169
  equimodular smallest, 165
  largest, 153,154,159
  localization problems for, 148
  multiple, 8,13,50,153,154, 169,181,221,247,267,322
  multiple dominant, 185
  near equimodular, 169,174
  near multiple, 169
  simple, 158
  simultaneous determination of all, 77,78
  smallest, 153,154,158,165, 166
  well-conditioned, 64
Zonneveld, J.A., 48
Zurmühl, R., 164

Index

uncertainty, 78,80,82,90,92,
 109
underflow, 7,21,52,54,55,57,
 58,60
unified convergence and
 estimation theory, 285
uniform convergence, 17,79,90
unitarily linearly convergent,
 100
unilateral polynomial equation, 22n
unsolvability), 115

V

van Zijlgaarden, T.A., 48

W

Ward, J.A., 195,196,197
Watkins, R.O., 172
Weierstrass, K., 178
Weyl, H., 77,78,80
Wheeler, D.J., 50
Wilkinson, J.H., 48,86/86,
 65,157,153,169,173
wrongal identifier, 136

Z

Zeros(s) (see also Polization)
 counting of, 131
 close, 106
 closest of, 8,13,181,190,
 211,247,282
 equimodular, 189
 equimodular smallest, 195
 largest, 152,154,159
 localization problems for,
 148
 multiple, 8,13,50,155,154,
 163,181,227,247,267,282
 multiple dominant, 285
 near equimodular, 193,174
 near multiple, 180
 simple, 156
 simultaneous determination
 of all, 73,74
 smallest, 157,156,158,165,
 166
 well-conditioned, 86
Zondervelt, T.A., 48
Zurmühl, R., 189